PROBABILISTIC TECHNIQUES IN EXPOSURE ASSESSMENT

A Handbook for Dealing with Variability and Uncertainty in Models and Inputs

PROBABILISTIC TECHNIQUES IN EXPOSURE ASSESSMENT

A Handbook for Dealing with
Variability and Uncertainty
in Models and Inputs

Alison C. Cullen
University of Washington
Seattle, Washington

and

H. Christopher Frey
North Carolina State University
Raleigh, North Carolina

Plenum Press • New York and London

Library of Congress Cataloging-in-Publication Data

Cullen, Alison C.
    Probabilistic techniques in exposure assessment : a handbook for
dealing with variability and uncertainty in models and inputs /
Alison C. Cullen and H. Christopher Frey.
        p.   cm. -
    Includes bibliographical references and index.
    ISBN 0-306-45956-6 (hc.). -- ISBN 0-306-45957-4 (pbk.)
    1. Environmental risk assessment--Mathematical models.
2. Probabilities.   I. Frey, Christopher H.   II. Title.
GE145.C85   1998
363.7'001'5118--dc21   98-44311
                                                               CIP

ISBN 0-306-45956-6 (Hardbound)
ISBN 0-306-45957-4 (Paperback)

© 1999 Society for Risk Analysis
Plenum Press is a division of Plenum Publishing Corporation
233 Spring Street, New York, N.Y. 10013

http://www.plenum.com

10 9 8 7 6 5 4 3 2 1

Printed in the United States of America

To those specific individuals (never random) most chronically
exposed to the spatial and temporal dimensions of this text:
the Cullen subpopulation (Chris, Ross, and Kyle)
and the Frey subpopulation (Caren and Lauren)

PREFACE

Five years ago, small groups of individuals hailing from government, the private sector, academia, and think tanks, united by their common desire to improve the application of probabilistic techniques in risk analysis, came together in a series of conversations. The application of probabilistic tools in risk analysis was in the midst of rapid evolution and there was concern about the broad availability of practical information and guidance. These individuals converged on the idea that a book of guidance and examples was needed to fill the gap. It was agreed that this document could be organized under the auspices of the Society for Risk Analysis (SRA), but should be informed by input and support from as many interested parties as possible. First one, and then two, young and unsuspecting authors were recruited to the cause.

A framework was conceived for developing the scope and coverage of the manuscript. This included extensive interviews with experts representing the many strands of environmental science and risk assessment interwoven in this area, a workshop involving topical experts to review an initial draft, and two rounds of extensive peer review as the text was edited and re-edited. The goal was to pull together relevant tools from a range of fields and sources, some easily located and some buried, and to illustrate their application, and in some cases extension, to meet the needs of exposure assessors pursuing probabilistic analysis. The current book is the result of these early steps and the liberal contributions of ideas, time, encouragement, and criticism of countless individuals.

Looking back it is difficult to know where to begin in acknowledging debts of gratitude to those who have taken part. The project certainly would not have been possible without the financial support of the American Industrial Health Council, the U.S. Department of Energy, General Electric, and the Harvard Center for Risk Analysis. In addition the project benefited from in-kind contributions from Alceon Corporation, SENES Oak Ridge, North Carolina State University, and the National Research Center for Statistics and the Environment. Financial oversight was handled by the SRA and a Management Committee comprised of: James Wilson (Resources for the Future), Richard Burk (SRA), Timothy Barry

CHAPTER 1

INTRODUCTION

"The logical method and form flatter that longing for certainty and for repose which is in every human mind. But certainty generally is illusion, and repose is not the destiny of man."

Oliver Wendell Holmes

Environmental regulators routinely make decisions about the extent to which human exposure to chemical contaminants will be controlled and/or reduced. These decisions commonly are aimed at protecting a targeted group, for example, the general population of a country, a highly-exposed subpopulation, or specific individuals with some characteristic in common, such as children. Individuals within a targeted group are likely to face different levels of exposure or dose due to differences in behavioral or dietary patterns (e.g., amount of time spent at work versus home or mass of home-grown tomatoes consumed each day), or physiological characteristics (e.g., breathing rates), or even over time. These differences lead to *variability* among individuals. Thus, to characterize exposures to an entire population, or any group or single member of the population, requires an understanding of the variability in exposure. In addition, one must understand and account for *uncertainty* about true levels of exposure. Uncertainty about exposure arises when there is a limited availability of empirical information, as well as imperfections in the instruments, models, or techniques used to develop representations of complex physical, chemical, and biological processes. For these reasons, estimates of emission rates, fate and transport efficiencies, human exposure factors, dose/response relationships, and the costs and benefits of regulatory options are often fraught with uncertainty.

If major decisions are to be adequately supported, variability and uncertainty should be explored as part of exposure assessment. In a broad-brush sense, the characterization of variability is obtained based on knowledge, whereas the characterization of uncertainty is intended to reflect the extent of ignorance about the true value of a quantity or the true distribution that represents variability. Ultimately, both the implications of the variation in exposure across populations and the degree of uncertainty associated with estimates of individual and population exposure must be considered during the decision making process.

This book contains a straightforward presentation of tools for addressing variability and uncertainty in exposure assessment. It is intended to address the unique needs of exposure assessors and risk managers in the context of human and environmental health decisionmaking. Approaches to developing frequency (or probability) distributions to represent variability (or uncertainty) in the inputs to exposure models, the propagation of these input distributions to generate single or multidimensional distributions of exposure, and the identification of significant contributors to overall variance in exposure, are presented. The book also addresses a number of other issues that exposure assessors need to consider, including uncertainties due to model formulation and lack of validation, and methods for iterating and refining an analysis.

1.1. BACKGROUND

Probabilistic exposure assessments are carried out for a diverse set of environmental hazards, both naturally occurring and anthropogenic. We acknowledge that variability in and uncertainty about toxicological factors frequently dominate the overall variance in probabilistic risk assessments (Bogen, 1990; McKone and Bogen, 1992; Cullen, 1995); however the broad application of probabilistic techniques to dose-response relationships in a regulatory context has not occurred. A handful of applications of probabilistic analysis to toxicity assessment is available in the published literature (Crouch and Wilson, 1981; Wilson and Crouch, 1987; Evans et al., 1994a&b). We recommend these to readers whose interests extend beyond the scope of this text. Still, we believe many of the issues discussed here in the context of exposure assessment will also be relevant to the characterization of uncertainty and variation in dose/response relationships. Finally, we note that many of the tools presented here are borrowed or adapted from other fields, especially the analysis of engineered systems, such as flood safety and control, structural design, and nuclear power technology.

Exposure models combine information about the frequency, intensity and duration of human contact with environmental contaminants and/or radionuclides through inhalation, ingestion, and dermal absorption. The model inputs can be categorized according to the roles they serve. Some describe the degree of contamination of various environmental media, others pertain to the behaviors, activities, or demographics of exposure and exposed populations, and still others describe human physiology. The terminology adopted by environmental health researchers, and used in this text, partitions exposure model inputs into those that are variable, those that are uncertain, and those with some aspects of each (Bogen and Spear, 1987; IAEA, 1989; Bogen, 1990; Finkel, 1990; Morgan and Henrion, 1990; Frey, 1992). In formal analysis, variability and uncertainty are characterized by two different probability spaces (Helton, 1996; Helton *et al.*, 1996).

Variability refers to temporal, spatial, or interindividual differences (heterogeneity) in the value of an input. Variable quantities include the rate at which individuals consume specific dietary items and the body weights of those individuals. Both of these quantities are also variable over time for a particular individual. Clearly the degree of variability in any quantity is influenced directly by the averaging time, geographic area, or other characteristics of the population under consideration. In general, variability can not be reduced by additional study or measurement.

Uncertainty may be thought of as a measure of the incompleteness of one's knowledge or information about an unknown quantity whose true value could be established if a perfect measuring device were available. For example, the dispersion factor used to estimate ambient air concentration in a particular location from meteorological and stack emissions data is uncertain due to our reliance on simplified models that describe the fate and transport of the emitted plume. Random and systematic measurement errors, as well as reliance on models or surrogate indicators, are all sources of uncertainty.

Variability and uncertainty also have different ramifications for decision-makers. "Uncertainty forces decision-makers to judge how probable it is that risks will be overestimated or underestimated for every member of the exposed population, whereas variability forces them to cope with the certainty that different individuals will be subjected to risks both above and below any reference point one chooses" (NRC, 1994). In confronting variability and uncertainty a decision-maker stands to better understand the degree of variance in the full distribution of exposure or risk, the impact of various assumptions, data gaps, and model choices on decision-making, and the most fruitful avenues for further study.

Throughout the history of risk assessment, (at least as performed in the context of facility siting, evaluation of emissions, and environmental cleanup), analysts traditionally have performed deterministic calculations incorporating point estimates representing the inputs to mathematical expressions of exposure and risk. A number of limitations of this approach have attracted increasing concern from a broad spectrum of interested individuals including regulators, the regulated community, the public, and environmentalists (Morgan and Henrion, 1990; Finkel, 1990; Burmaster and Harris, 1993; Cullen, 1994; Thompson and Graham, 1996). First the degree and direction of bias or conservatism (economic or environmental) may be completely masked by the reporting of a single number. Further, a deterministic analysis does nothing to guide decision-makers about whether to conduct additional research or select from available options to reduce exposure. Where large stakes decisions ride on exposure analyses, these limitations may not be acceptable.

These factors have led to increasing interest in probabilistic techniques, where variability and uncertainty are represented explicitly. The richness of the analysis and the potentially revealing results are attractive, but the information needs can be demanding, leading analysts to an iterative framework in which problems are more fully explored and understood as appropriate. Existing texts and papers provide excellent coverage of some of the available tools developed in the fields of uncertainty analysis, probabilistic analysis, and decision analysis (IAEA, 1989; Bogen, 1990; Finkel, 1990; Morgan and Henrion, 1990; Frey, 1992; USEPA, 1996b). However, there remains a need for straightforward guidance about these tools (and extensions of these tools), their utility and limitations, and specific examples of their use in the development of probabilistic descriptions of environmental factors. Hence this text has the following goals:

- To provide guidance on how to develop probabilistic descriptions of inputs that reflect variability, uncertainty, or a combination of both, particularly in cases where some data are available to support the development of distributions;
- To provide guidance on how the implications of variability and uncertainty may be assessed, particularly in terms of identifying key factors as a means for setting research or data collection priorities;
- To provide guidance on how to identify and address uncertainty in models.

Decision-makers benefit from probabilistic information in several ways. Characterization of uncertainty allows decision-makers to choose whether to actively reduce exposure or to conduct additional research. Also, assessments of uncertainty in exposure estimates can be used to prioritize opportunities for additional scientific research that would address resolvable sources of uncertainty. Furthermore, characterizations of variability and uncertainty in exposure under alternative control scenarios enable more rigorous comparisons than the use of upper-bound point-estimates would. In fact, an upper-bound exposure under the assumptions of one control strategy may be far more plausible than an upper-bound under the assumptions of a different control strategy (HCRA, 1994; Thompson and Graham, 1996).

1.2. HISTORICAL CONTEXT FOR PROBABILISTIC EXPOSURE ASSESSMENT

The history of the use of probabilistic techniques is long and varied. The versatile tools of uncertainty analysis are mentioned in accounts of the American Civil War, and were a topic of interest early in the twentieth century in British Universities (Hammersley and Handscomb, 1964). Over time the approaches have become more sophisticated in response to the needs and uses to which they have been applied. In a major advance during the 1940s, Monte Carlo simulation, a random sampling technique for solving difficult deterministic equations, originated at Los Alamos from the work of Ulam, von Neumann, and Fermi (Hammersley and Handscomb, 1964; Ulam, 1976; Rugen and Callahan, 1996). Introduction of explicit techniques for the treatment of uncertainty in major United States regulatory policy decisions has occurred over the past 20 years or so. One of the first examples of this was the United States Atomic Energy Commission sponsored "Reactor Safety Study" (NRC, 1975) which together with reviews and follow-up studies led the Nuclear Regulatory Commission to develop probabilistic safety objectives. In addition, the field of engineering has a long history of uncertainty analysis applied to safety and cost estimation, especially for power plants and chemical plants.

Meanwhile, exposure and risk analysis were evolving to a point where in 1983, the National Research Council presented the basic concepts for use in regulatory settings through a seminal report often referred to as the "Red Book" (NRC, 1983). William Ruckelshaus, then administrator of the United States Environmental Protection Agency (EPA), championed the new ideas and methods at the agency. Through his leadership, the EPA began to use risk analysis in rulemakings and in the nascent Superfund program.

The early guidance documents released by EPA for use in Superfund site assessments and cleanups assumed deterministic representation of all but a few exposure model inputs. However, in its 1989 "Exposure Factors Handbook" the agency included deciles for exposure variables supported by large data sets, such as body weight, along with recommended point estimates. In recent years, the focus of technical leadership in probabilistic exposure assessment has been shared by the United States Department of Energy (DOE), the U.S. Department of Defense (DOD), the EPA and many others, leading to a host of innovations which continue to appear in federal risk assessments. For example, EPA's Office of Air Quality Planning and Standards (OAQPS) has a substantial track record in the application of probabilistic techniques to the development of air quality standards (Johnson et al., 1993; Whitfield and Absil, 1994). Also, the DOE has supported the development of new probabilistic analysis and optimization methods, with applications to technology assessment and environmental remediation (Frey and Rubin, 1992a,b,&c).

There are a number of institutional motivations which are converging to prompt analysts at federal agencies to characterize variability and uncertainty in their assessments quantitatively, and to communicate this information to risk managers. These include the study by the National Research Council entitled *Science and Judgment in Risk Assessment* (NRC, 1994), evolving practice in academia and industry, requests for uncertainty analyses by the Office of Management and Budget (OMB), memoranda on risk characterization representing formal agency guidance crafted by EPA's Risk Assessment Commission and

Deputy Administrator F. Henry Habicht, II (USEPA, 1992a) and Administrator Carol Browner (USEPA, 1995a), and discussions of risk and exposure assessment taking place around a series of bills proposed by members of the 103rd, 104th, and 105th Congresses.

Most recently a memorandum approving probabilistic risk assessments for agency use was issued by EPA's Science Policy Council (USEPA, 1997b). EPA's position is that probabilistic tools are acceptable as long as analysts adhere to certain principles. The agency has emphasized the importance of clarity, transparency, reasonableness, and consistency in risk assessment. To these principles they now add a number of specific conditions for acceptance of Monte Carlo analyses which appear in Table 1.1. We note that these conditions are sound suggestions for any risk assessment, not just one involving probabilistic approaches.

In summary, we note that throughout its history, probabilistic analysis has been subjected to intense scrutiny from all quarters. Probabilistic analysis places emphasis on developing model input assumptions based on all available information and knowledge. This process requires "laying on the table" all relevant information for consideration. In contrast, point-estimate analyses require decisions about what to exclude from the analysis: only a single number can be used to characterize each input and, therefore, many other possible values of that input are ignored. In many instances, someone reviewing a point-estimate analysis may have little information about how the point-estimates are chosen; however, probabilistic analyses, which tend to be more explicit about what is and is not known, sometimes attract more criticism. Unfortunately, concerns may arise that probabilistic approaches are being used as a smoke screen to confuse deci-

TABLE 1.1 Conditions for Acceptance of Monte Carlo Analyses by USEPA

1. The purpose and scope of the assessment should be clearly articulated in a problem formulation section.
2. The methods used for the analysis (including models, data, and assumptions) should be clearly documented and easily located in the report.
3. The results of sensitivity analyses should be presented and discussed in the report.
4. The presence or absence of moderate to strong correlations and dependencies between input variables should be discussed and accounted for, along with the effect these have on the output distribution.
5. Tabular and graphical representations of the distributions of input and output including statistics and percentiles of special interest should be provided. Variability and uncertainty should be differentiated.
6. The numerical stability of the central tendency and higher end (i.e., tail) of the output distribution should be investigated and discussed.
7. Deterministic calculations should be provided for comparison with Monte Carlo output.
8. The consistency and appropriateness of the metric used for the exposure estimates, and that of available toxicity information, should be discussed.

Source: (USEPA, 1997)

sions or issues when these tools are introduced in an adversarial process. None of this should discourage the use of probabilistic methods, but rather, it should incite their cautious and judicious application.

1.3. WHEN IS PROBABILISTIC ANALYSIS USEFUL AND/OR JUSTIFIED?

The protection of public health from environmental contaminants was (and remains) the goal of much of the regulation relying on these assessments. Thus conservative point estimates of variable and uncertain inputs were sought to ensure "worst case" or upper-bound estimates of risk and exposure. Such worst case analyses may be very valuable because they can be quick, cheap, and can serve to identify potential exposures that are so low that detailed analyses are not warranted. Of course, due to variability in and uncertainty about exposure, the degree and direction of the conservatism associated with deterministic inputs and outputs is unknowable without a detailed description of the specific exposure scenario and the receptors under consideration. Even with such information, the degree of conservatism may be difficult to establish. Despite these limitations, deterministic point-estimates sometimes enjoy a precise and/or accurate *appearance*, and inspire a misleading sense of confidence. Further, deterministic estimates based on conservative inputs provide no indication of the magnitude of uncertainty surrounding the quantities estimated and lend no insight into the key sources of underlying uncertainty (although some insight can be provided with deterministic sensitivity analysis). Also, when several outcomes are of interest the term "conservative" may become ambiguous. In other words, inputs that lead to conservative estimates for one outcome may not lead to conservative estimates for another.

In contrast to the point-estimate approach, probabilistic analysis yields a fuller characterization of the information and state-of-knowledge which affect a decision. With quantitative information about both the *range* and *likelihood* of possible exposures for a set of individuals, decision-makers can assess whether a particular decision is likely to be robust to variability and/or incomplete knowledge. In addition, probabilistic methods can help increase the likelihood of success for improving the knowledge base with the most pertinent information.

In recent years exposure analysts have begun to recognize the power of the probabilistic approach. As with any tool there are circumstances in which it can be meaningfully applied and circumstances in which it can not. Some exposure assessors are torn between an admiration of the possibilities of probabilistic analyses and a concern that the distributions assigned to model inputs be based on "facts" (USEPA, 1992a). Certainly, the use of distributions based on empirical evidence is desirable; however, some inputs have not been, and/or can not be, measured under the conditions of interest. This is especially true of attempts to

TABLE 1.2 Cases in Which Probabilistic Analyses May Be Useful

- When the consequences of poor or biased exposure estimates are unacceptably high
- When a screening level deterministic calculation indicates exposures of potential concern, but carries a level of uncertainty that does not warrant immediate expenditures on remediation
- When there is interest in the value of collecting additional information, such as when time and resources permit additional sampling, but questions remain about whether this will improve the quality of the decision to be made
- When uncertain information stems from multiple sources
- When significant equity issues are raised by sources of variability, such as when subpopulations face unusual exposures relative to those of the general population
- When assessing the potential benefits of targeting resources for various interventions, for example when more than one strategy for remediation is available, but one would reduce exposure via the food chain while another would improve air quality
- When ranking or prioritizing exposures, exposure pathways, sites, or contaminants is important
- When the cost of remedial or intervention activity is high

predict future exposures. In these cases, regardless of whether one is pursuing deterministic or probabilistic calculations, judgment must be applied. Table 1.2 identifies some situations in which the benefits of probabilistic analysis are compelling. Conversely, Table 1.3 identifies situations in which a probabilistic analysis may be unwarranted.

Arguments in favor and against probabilistic analyses and deterministic analyses in specific scenarios are discussed further by others (Hattis and Burmaster, 1994; Pate-Cornell, 1996; USEPA, 1997b). Overall the benefits of probabilistic analysis depend on the nature of the problem and the needs of the decision-maker or risk manager.

1.4. ASSESSING THE EXISTING INFORMATION BASE

The decision to pursue empirical and/or subjective approaches to the development of an exposure model input distribution also must be sensitive to the quantity and relevance of the available information about that input. Inputs can be partitioned on the basis of what is known about them, as illustrated in Fig. 1.1.

TABLE 1.3 Cases in Which Probabilistic Analyses May Not Be Useful

- When a screening-level deterministic calculation indicates that exposures are negligible,
- When the cost of averting the exposure is smaller than the cost of probabilistic analysis,
- When safety is an immediate and urgent concern,
- When probabilities are so uncertain and/or indeterminate that detailed probabilistic judgments are impossible,
- When there is little variability or uncertainty in the analysis.

Data Quantity

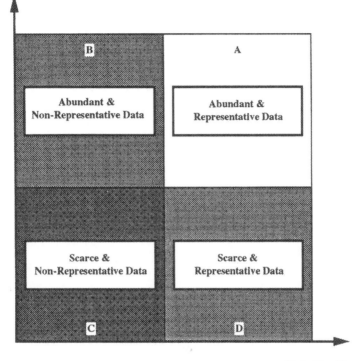

Data Quality

FIGURE 1.1 Categorization of data based on quality and quantity.

When empirical data are plentiful, and relevant to the exposure of concern, distribution development may proceed using statistical techniques (Box A of Fig. 1.1). Human physical characteristics, such as body weight and height, fall into this category. Alternatively, inputs characterized by a complete absence of empirical data are candidates for distributions developed using subjective approaches (Box C of Fig. 1.1). Such inputs include quantities which must be represented for a point in time some decades or centuries in the future (though even in these cases clues derived from present experience may be found). An evaluation of how many people will visit an as yet unbuilt park is an example of such an input. Inputs for which directly relevant data are unavailable, but for which data obtained in alternative locations, populations or time frames exist, are candidates for a mixed approach in which judgment is used to adapt the distribution based on nonrepresentative data, such as to adjust its moments to reflect the scenario of concern (Box B in Fig. 1.1). We acknowledge that assessments about the relevance of available data are largely judgment-based. An example of such an input is the dietary consumption rate for a specific food by a specific unsurveyed population (while survey data for other foods or populations are available). The final category consists of inputs for which scarce but relevant data are available. These inputs are candidates for a statistical approach in which attention is paid to addressing uncertainty in the parameters of the distribution resulting from the small sample size (Box D in Fig. 1.1). The development of a distribution to represent the annual average concentration of a contaminant in air based on a small number of measurements spread across the seasons would fall into this category. In our experience, many inputs are in categories B and D since empirical information is often: (a) unsatisfactory with respect to specific questions in a given assessment; (b) limited or nonexistent; or (c) impossible to obtain.

1.5. STRUCTURE OF THE TEXT

In accordance with our stated goals, we have attempted to illustrate in a straightforward manner the application of tools from the field of uncertainty analysis to the practice of probabilistic exposure assessment. The flowchart appearing as Fig. 1.2 provides a broad outline to probabilistic exposure assessment and the structure of this book. The interlinking arrows are intended to represent the iterative nature of probabilistic analyses, as well as the interrelationship of virtually all of the components.

While many analysts will come to this text with a well-defined problem, in Chaps. 1, 2, and 9 we discuss the importance of identifying/establishing the source and nature of contamination, the population exposed, the averaging time of the exposure, and very important, the goal of the analysis. Chaps. 2 and 3 are concerned with model and input uncertainty, the implications of selecting individual models, the consideration of alternative models and model forms, the relationship of the input uncertainty and the model uncertainty, and approaches to addressing model uncertainty. Chaps. 4, 5, and 6 look at approaches to analysis and assessment of variability and uncertainty in model inputs and Chap. 9 presents specific examples. Chap. 4 examines the theoretical basis for distributional assumptions along with decisions about when and why to pursue probabilistic descriptions of inputs to exposure models. Chap. 5 covers data analysis approaches for establishing representative distributions, including uncertainty in statistics, the empirical basis for selecting distributions, parameter estimation methods, and small data sets. In Chap. 6 we look at special topics including maximum entropy inference, combining data, default or initial distributions, surprise, and dependence between model inputs. We also introduce expert elicitation very briefly. Chap. 7 discusses methods for propagating distributions through models based

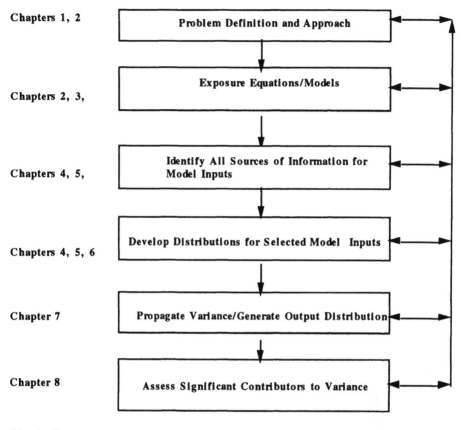

FIGURE 1.2 Components of probabilistic uncertainty analysis with pointers to book sections.

upon fundamental theorems, analytical solutions, and numerical simulations, as well as methods for two-dimensional simulation of variability and uncertainty. Chap. 8 focuses on the identification of the most significant contributors to variance in the result (introducing the concept of "value of information"), and the iterative development of distributions. Finally, a case study appears as in Chap. 9.

1.6. COMPOSITION OF EXPOSURE ASSESSMENT TEAMS

A few words are required regarding the composition of teams needed to conduct rigorous exposure assessments. Individuals encompassing a range of skills are required to pursue distribution development and probabilistic exposure assessment in general. What type of team should be formed? Obviously the answer depends on the nature of the exposure or dose under consideration and the decision to be faced. First, the risk manager or decisionmaker, whose information needs must be served by the analysis, is integral to the team. Stakeholder groups and members of the general public should be included to help ensure that the analysis is acceptable and addresses their concerns. Individuals familiar with the agent or environmental contaminant of concern are essential, such as scientists, engineers, and/or plant managers. Also, for the adequate treatment of exposure scenarios in which the fate and transport of contaminants adds to the analytical complexity, it is necessary to include a fate and transport model expert and/or environmental chemist. A statistician and/or an analyst experienced in probabilistic modeling may be helpful, particularly one specializing in the type of data relevant to the analysis. A behavioral scientist is an important element of the team if time/activity or other behavioral survey data will be analyzed. Finally, an experienced practitioner in expert elicitation is a critical source of guidance for both formal and informal elicitation of knowledge, belief, and experience surrounding the adaptation and use of measured data.

1.7. CONVENTIONS USED IN THIS TEXT

In this section we define terms and conventions used throughout the text, as well as the intended scope.

1.7.1. Terminology

Exposure assessment is inherently a multidisciplinary activity. Thus, a potential barrier to the clear communication of ideas is the lack of a common language across fields. We strive to make explicit our own use of terminology. We have included a glossary of terms to prevent confusion.

This text focuses on approaches to describing variability and uncertainty in inputs to exposure models and the models themselves. Exposure model inputs include quantities such as the concentration of a contaminant in an environmental medium, physical characteristics (body weight, skin surface area, etc.), and exposure descriptors and behavioral preferences (e.g., consumption rate of local foods, duration of residence in a particular location, etc.). These quantities are referred to by some as "exposure factors" and "contact media concentrations," by others as exposure model "parameters" or "variables," and the list of alternative terms goes on. In this text these quantities will be referred to as exposure or dose model inputs, or simply *inputs*. The results or end points generated by model calculations will be referred to as exposure or dose model output distributions, or simply *outputs*. Units are provided for all terms and calculations. For consistency, references to logarithms indicate natural logarithms unless otherwise noted.

1.7.2. Use of Examples

To provide a concrete demonstration of some of the concepts and techniques presented we include brief examples throughout the text. Also, we refer to many published examples from exposure and risk assessments to further illustrate concepts and calculations. Finally, Chap. 9 contains a somewhat lengthy illustrative case which demonstrates a series of key concepts in a single example.

1.7.3. Scope

Due to space limitations and our intention to raise a broad range of concepts, and to include the work of many researchers, we rely on extensive referencing to cover as much material as possible. These references include primary literature, text books, agency publications, and readily available opinion pieces. A full and complete understanding of these works requires substantial background in a diverse set of fields. We hasten to note that the present text is in no way a substitute for education or experience in probability and statistics tempered with large doses of common sense, curiosity, and humility. These pages represent a compilation of information and advice from many sources that we hope will serve to guide analysts seeking to produce sound and applicable distributions for use in probabilistic exposure and risk assessments.

1.8. SUMMARY OF GOALS AND PHILOSOPHY

Our philosophical framework in pursuing the three goals presented in Section 1.1. is to encourage analysts to recognize and address uncertainty explicitly in all facets of a probabilistic assessment. We assume that the majority of our readers are working on simulated or analytic solutions for complex exposure models; however, the broader applicability of many of these tools is certainly acknowledged. Quantitative approaches are suggested, along with guidance on how to gauge their applicability under a number of conditions. In addition, we would encourage analysts not to oversell or undersell their findings. It is critical to be honest and open regarding the limitations and applicability of any analysis, as well as one's level of confidence in the results. Finally, it is our hope that the tools for interpreting results suggested here will foster improved communication with decision-makers.

CHAPTER 2

A BASIC FRAMEWORK FOR PROBABILISTIC ANALYSIS

In this chapter, we consider four major issues. The first includes two of the philosophical frameworks by which analysts conceive of probability. This discussion is important because the types of analyses and information used by analysts depend upon these philosophies. The remaining issues come under the general topic of a taxonomy for variability and uncertainty. These three are model uncertainty, variability in model inputs, and uncertainty in model inputs. This chapter will categorize, define, and discuss them. The later chapters of this book will show how to deal with them.

2.1. PHILOSOPHY OF PROBABILISTIC ANALYSIS

Hahn and Shapiro (1967) describe two of the major interpretations of probability. These are: (1) a frequency (or empirical) interpretation of probability; and (2) a subjective interpretation of probability. Some analysts adhere strictly to the first view. For many others, the subjective interpretation is preferred in cases for which there are few data and/or when there is a lack of representative data. Understanding these viewpoints can help identify the sources of potential differences in opinion among analysts about how best to develop probability models for inputs to an exposure or risk model.

2.1.1. The Frequentist View of Probability

Suppose that one conducts an experiment in which values are measured. Due to errors in the measurement process, there is a random distribution of values even if the quantity which is being measured is a fixed but unknown constant. For example, suppose that the true value for a concentration is 10 $\mu g/m^3$, and that we employ a sampling and analytical procedure with an unbiased, normally distributed error of ±10% to repeatedly measure this concentration a large number of times. Assuming that the error range from

−10% to +10% represents a 95% probability range, our set of observations would be a normal distribution with a mean of 10 μg/m³ and a standard deviation of approximately 0.5 μg/m³. This is because the 95% probability range is enclosed by the 2.5 and 97.5 percentiles of the distribution, which are ±1.96 standard deviations from the mean. If we took a large number of samples (e.g., 1000) then we could group the samples in bins of some standard width (e.g., in intervals of 1 μg/m³). The percentage of the total number of samples which would fall into any of the bins would represent the frequency with which we obtained samples within the specified range of values (e.g., from 8 μg/m³ to 9 μg/m³). As we increase the sample size, the frequency of obtaining samples within a particular range of values will stabilize. The frequency with which we obtain samples in a specific range is referred to by frequentists as the *probability* of the event or set of outcomes. Thus, in this case, the probability that we obtain sample values less than 10 μg/m³ is 0.50 because, as we go to very large sample sizes, half of the samples will be less than the true value. As another example, the probability that we would obtain values between 8 μg/m³ to 9 μg/m³ in the shaded region of Fig. 2.1 is 0.0224.

In this example, we have relied on a hypothetical empirical data set. The frequentist view requires that we base statistical inferences upon empirical data. This view is therefore often referred to as the empirical approach. Furthermore, it is preferred that the data be collected at random from a defined area or population. Thus, if we have no data, and are true frequentists, we cannot say anything about the probability distribution for a quantity. In this view, inferences about the values of parameters (e.g., mean, variance) of probability distributions must be based upon empirical data.

The frequentist approach is also sometimes referred to as the "classical" approach, because it is based upon long-established principles of sampling theory. However, DeGroot (1986) points out that the "classical" interpretation of probability is somewhat different from the frequentist approach. In

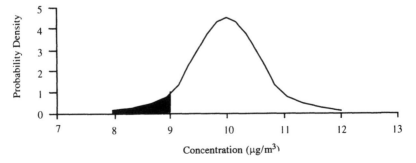

FIGURE 2.1 A distribution of unbiased measurements for a fixed unknown quantity.

the classical approach, the concept of probability is derived from fundamental theory and thought experiments, such as regarding a toss of a coin, roll of a die, or draw of a card from a deck.

One purpose of statistical inference, regardless of the underlying philosophy, is to develop compact representations of data sets. For example, suppose that we have 20 measurements. We can work with all 20 measurements, or we can use methods of statistical inference to estimate the values of the parameters of a probability distribution model that is a good representation of the dataset. Thus, instead of having to work with 20 data points, we may be able to represent the data with only two parameters (e.g., a mean and variance) and an assumed probability model (e.g., a normal or lognormal distribution). Thus, the power of statistical inference is to develop compact and convenient models for representing data. It is important to remember that a probability distribution is in fact a type of model. A probability distribution allows one to make interpolations between observed data points. Furthermore, it is a basis for extrapolating to the tails of the distribution, which are rarely well-represented in finite data sets.

Box and Tiao (1973) discuss an example in which 20 observations are available. The information in the 20 observations can be summarized in terms of inferences about two parameters, such as the mean and variance, conditional on the assumption of a particular probability model, such as a normal distribution. Thus, given estimates for the mean and variance of the dataset, one can do statistical analysis to evaluate whether, for example, a normal distribution can be rejected as a poor fit to the data set. This type of analysis is discussed in Chap. 5. If a normal distribution appears to be a good representation of the data, then inferences can be made regarding the statistical confidence intervals for the parameters of the distribution.

The inferences which can be made about the data set depend upon what probability model is assumed. For example, if the data are highly skewed toward large values, a normal probability model may not be appropriate. The use of the normal distribution as a probability model in this case could lead to distortion of information about the extreme values of the data set.

Statistical analysis should be an iterative process. Given a data set, the typical first step is to postulate a probability model (e.g., normal, lognormal, beta, gamma, etc.). Then, an effort is made to build and test the model. This typically involves using statistical estimators for the parameters of the probability models (distributions). An estimator is a mathematical relationship between a data set and a parameter of a distribution. The statistical inferences made using the estimators will depend on the adequacy of the assumed probability models. For example, if we are trying to calculate the shape parameters, α_1 and α_2, for a beta distribution, the values we obtain for α_1 and α_2 will be meaningless if there is no shape of a beta distribution that can reasonably represent the data set. Thus, up to the point where we estimate the values of the parameters, we behave as if we believe the model to be correct. Once we obtain the parameter values, we can compare the distributional model to the data set. In this step, the adequacy of the model is critiqued. This is often done using goodness-of-fit tests. We might consider, for example, the nature of the residual error between the observed data and the predictions that we would make from the probability distribution model.

We will consider some of the details of statistical inferences based upon a frequentist view in later chapters. Here, we are primarily interested in drawing out the philosophical distinctions between the frequentist and subjectivist views of probability.

2.1.2. Subjective or Bayesian Views of Probability

A frequentist who has no data is paralyzed. However, it is rarely the case that nothing is known about an input to a model even in cases for which data are lacking. Typically, it may be possible to make analogies with similar quantities as a basis for selecting an appropriate probability model to represent variation in the quantity. For example, as discussed in Chap. 4, there is a theoretical basis upon which to select distribution types in many cases. It is often possible to place upper and lower bounds on the possible value of the quantity due to physical or practical limitations. In the case of an exposure model, inputs such as body weight or concentrations must be non-negative. Furthermore, such quantities are not boundless. For example, it is rare to find humans who weigh more than 1000 lbs. Thus, even in the absence of site-specific data for a particular exposure assessment, it is often practical to make judgments about the range of values and shapes of probability distributions that can be used to describe model inputs.

Even in cases where data are available, the data may not be representative of or relevant to the problem at hand. Thus, statistical evaluation of data may be an insufficient basis for estimating variability or uncertainty in a quantity. As a result, some degree of judgment about the available data is required. Furthermore, even the application of statistical techniques, such as goodness-of-fit tests, requires considerable judgment and experience, as is discussed in Chap. 5. For example, the analyst makes judgments about what types of parametric distributions are appropriate to represent uncertainty in a given empirical quantity, even though the analyst may rely on the data and the goodness-of-fit test to make inferences regarding the values of the distribution parameters (e.g., mean, variance) and to critique the selected probability model.

Thus, there may be cases in which data are lacking in quantity or quality but for which an analyst has other information that can be used to construct a probabilistic representation of an input to a model. This alternative to the frequentist approach is often referred to as a "subjectivist" approach. A rigorous theoretical basis for this approach has developed which in many ways parallels the frequentist approach. The theoretical basis is known as Bayesian inference. The key feature of the theory of Bayesian inference is Bayes' Theorem, which was first published in 1763 (Box and Tiao, 1973).

Bayes' Theorem involves the use of a so-called "prior" distribution, which is intended to express the state of knowledge (or ignorance) about the parameters of a probability model. The prior distribution can be developed on purely subjective grounds in the absence of data. For example, while for some reason it may not be possible to measure fish consumption rates for a population of subsistence fisherman, it may be possible to make inferences about that distribution based upon analogies with other data sets. One might be able to place upper and lower bounds on the fish consumption by assuming that there is variation regarding how many meals per day are based upon fish, how much fish would be consumed in a single meal by a person of a particular body weight, and so on. For simplicity, let us suppose that the logarithms of fish ingestion rates are normally distributed (i.e. fish ingestion is lognormally distributed). A prior distribution, in the absence of data, would be specified by a judgment for the mean and variance of the logarithms of fish ingestion rates.

The power of Bayes' Theorem is in combining a judgment with data that are subsequently collected. The process of combining data with a prior distribution is called updating. Bayes' Theorem specifies the mathematics for such a combination, which results in the so-called "posterior" distribution. The mean of the posterior distribution is typically some type of weighted average of the mean of the prior distribution and of the observations in the newly obtained data set. The variance of the posterior distribution is also based upon a combination of the variance of the prior and of the observations. As the number of data points becomes large, the influence of the prior distribution will wane, and the results from the Bayesian updating method will approach those of a traditional frequentist analysis of the same data set. DeGroot (1986), Box and Tiao (1973), and others provide more detail regarding Bayes' Theorem.

The intermediate results obtained from Bayesian inferences depend on the nature of the prior assumption. Consider two analysts with differing states of knowledge about a quantity. Analyst A, who may have done many analyses with similar quantities, may generate what is termed an "informative" prior. This is a prior that has a relatively narrow range. In contrast, Analyst B may be more ignorant of the quantity, and would generate a much wider distribution. A small number of data points would add little to Analyst A's knowledge about the quantity, and thus A's posterior distribution may look very similar to the prior. However, those same data points may substantially increase B's knowledge, and therefore B's posterior may be significantly different. For both analysts, after many data points are obtained, their posterior distributions will be dominated by the information contained in the data, and their posterior distributions will converge toward each other if they are using the same data sets.

2.1.3. Discussion

The frequentist and Bayesian approaches are often pitted against each other and have been the subject of many debates. Our purpose here is not to resolve these debates, but to try to point out some similarities and differences in the approaches. Both approaches should give the same answer when there are plentiful, representative, and randomly drawn data. In some ways, the frequentist approach is more restrictive. This is because, in situations in which there are no data, it is not possible for a frequentist to perform a probabilistic assessment. A Bayesian would, in such a situation, work with a subjective prior distribution, which would be subject to updating if data were obtained. In cases in which relatively few data points are available, a strict application of sampling theory and frequentist principles may make it difficult to make meaningful inferences from the data set. For example, with only three or five data points, it is difficult to reject many probability models on the basis of being a poor fit to the data. The selection of a probability model must be guided by judgment in all cases, but in the absence of data, judgment may be relied upon more heavily.

In many cases data are not truly representative of the empirical quantity that an analyst desires to represent as a probability model, and the data may not be a random sample. Thus, biases may exist. In order to quantify biases, it is necessary to have a benchmark. In the absence of a benchmark, judgments may be used. For example, if fish ingestion rates are available for the general population, but must be estimated for subsistence fisherman, it would be reasonable to assume that the mean ingestion rate for the subpopulation of interest is greater than that for the general population. Failure to account for this fact would lead to substantial errors in the analysis.

The principal drawback of the Bayesian approach, often noted by those who prefer classical or frequentist methods, is the lack of consistency and replicability of inferences when comparing experts with different prior beliefs. Since expert judgments that are the basis for prior estimates are subject to biases for a number of reasons (e.g., Kahneman *et al.*, 1982), differences in opinion are expected. Furthermore, differences in opinion are likely to arise if experts do not share the same knowledge and experience. However, Bayesian analysis can be helpful in elucidating the influence of different prior assumptions relative to the information contained in available data, and in determining the amount and quality of data necessary for convergence to the same posterior distribution. Thus, Bayesian analysis can be helpful in designing data collection programs for situations involving stakeholders with widely differing prior beliefs regarding the risk associated with a particular exposure (e.g., Wolfson *et al.*, 1990).

The overall suggestion here is that the subjectivist or Bayesian approach, despite its drawbacks, is more flexible in dealing with the very common, practical situations in which data are limited. Furthermore, while the Bayesian approach explicitly acknowledges the role of judgment, the frequentist approach also requires judgment. The selection of a probability model by which to evaluate a data set is a key example of this. Certainly it is rare that any set of empirical data constitutes the full body of knowledge and information available and relevant to these quantities. In the absence of such information, Bayesian approaches are able to blend information from multiple sources of data in the quest for a representative distribution.

2.2. TAXONOMY OF VARIABILITY AND UNCERTAINTY

There are a number of distinct sources of variability and uncertainty in analyses of environmental problems. The latter come under the general headings of model or structural uncertainty and parameter uncertainty. Several authors, including Bogen and Spear (1987), Morgan and Henrion (1990), Finkel (1990), IAEA (1989), Frey (1992), NRC (1994), NCRP (1996), and others, provide more detail regarding frameworks for categorizing sources of uncertainty. Sources of uncertainty are also discussed in some U.S. EPA documents (e.g., USEPA, 1992b; 1996b; 1997a). A few key concepts are summarized here. First, we begin with a discussion of model uncertainty, followed by a discussion of uncertainty in model input variables. Model uncertainty is treated first because it is often overlooked and in some cases can be more important than input uncertainties. If a model is mis-specified and, therefore, cannot effectively characterize the desired output, then an uncertainty analysis of its inputs is meaningless.

2.2.1. Model Uncertainty

The structure of mathematical models employed to represent scenarios and phenomena of interest is often a key source of uncertainty, due to the fact that models are often only a simplified representation of a real-world system, and that the problem boundary encompassed by a model may be incomplete or incorrect. The problem boundary encompasses anything that is included in the model, which might be a geographic area, a time frame, or a population. The boundary may include only specific features pertaining to the spatial and temporal dimensions of the problem, such as a limited list of chemical or exposure pathways. Often, the problem boundary is referred to as a scenario in the context of exposure assessment.

Significant approximations are often an inherent part of the assumptions upon which a model is built. Competing models may be available based on different scientific or technical assumptions. Model uncertainty may be small for a well-tested physical model, or may be large for domains where the science is immature and data are lacking for model testing and validation. Furthermore, process, spatial, or temporal aggregation (e.g., grid size or averaging period) of many models is also a type of approximation that introduces uncertainty into model results.

In exposure assessment, there are a wide range of models employed. These models range from very simple multiplicative models containing only a handful of inputs to extremely complex structures. A classic example of a simple exposure model is one that is often used to calculate potential dose for chronic health effects, where risk is assumed to be linearly associated with dose:

$$\text{Dose} = \frac{\text{IR} \cdot \text{C} \cdot \text{ED}}{\text{BW} \cdot \text{LT}} \tag{2.1}$$

where

    Dose = lifetime average daily dose (mg of chemical per kg of body weight per day)

    IR = Intake Rate. For inhalation, this is typically expressed as liters per minute

    C = Concentration of chemical in environmental medium (e.g., micro grams per cubic meter of air)

    ED = Exposure duration (e.g., years)

    BW = Body weight (kg)

    LT = Lifetime (years)

This model is a highly simplified version of reality, in which dose, i.e., in this case the lifetime average daily dose (mg/kg/day), depends on the average concentration of contaminants in environmental media, average intake rates, body weight, exposure duration, and lifetime. The estimates of intake rate, concentration, and body weight are based on long-term averages. By using average values in the equation, an implicit assumption is that the health effects are independent of the time–history pattern of exposures.

When is this simplified version appropriate? The answer depends on the end-point of the exposure assessment. If the exposure assessment deals with a chronic health effect resulting from a long-term exposure when there is little temporal variability in the model inputs (especially for intake rate, concentration, or body weight), then this simple model may be appropriate. However, if the end-point is an acute health effect resulting from a short-term exposure, then the use of average concentrations and average intake rates would lead to errors in predictions. In such cases, a more detailed model should be employed that takes into account the temporal variation of the inputs (e.g., Price *et al.*, 1996a&b). The selection of averaging time should be consistent with the input requirements for the dose-response model to be employed.

Exposure assessments may also include models of different types of phenomena. Specifically, in determining the exposure of a population to air pollution emissions from an incinerator as in Cullen (1995), one needs models of how emissions are generated at the incinerator, how the pollutants are transported and transformed in the atmosphere, where the pollutants are deposited, how the pollutants are transported through various exposure pathways (e.g., food chain), what concentrations of pollutants exist in different media (e.g., air, food), and what quantities of media are consumed by members of the exposed population. An additional step may involve the use of human pharmacokinetic models to estimate the dose of chemicals to internal organs. Thus, an integrated exposure assessment may include a mix of engineering, physical, chemical, biological, and behavioral models. In some cases, mass and energy balances may be easily developed for well-defined systems. In others, lack of information about rate-based processes may hamper accurate predictions of chemical transformations through each step in a pathway. Thus, empirical factors (e.g., partitioning factors) may be employed in place of models with a theoretical basis.

Various sources of model uncertainties, and how they may be evaluated, are summarized in the following subsections. This discussion is based upon Frey (1992).

2.2.1.1. Model Structure. Alternative sets of scientific or technical assumptions may be available for developing a model. The implications of these various foundations may be evaluated by constructing alternative models and comparing the results from each. In some cases, it may be possible to parameterize model structures into a higher order model, and to evaluate alternative models using traditional sensitivity analysis. An example is a general formulation of a dose-response model to include or exclude a threshold and to have either linear or nonlinear behavior (e.g., Morgan and Henrion, 1990, p. 68). A threshold parameter can be set to zero to represent a nonthreshold model, and an exponent parameter can be set to one to represent a linear model.

As an example of parameterization in exposure assessment, the dispersion coefficients in an air quality model may be generalized in order to simulate a variety of atmospheric stability classes and other parameters may be included to account for the possibility of downwash or fumigation.

When there are alternative underlying assumptions between competing models, it may not always be possible to determine *a priori* which is more "correct." A typical approach is to report the key assumptions underlying each model and the corresponding results. If the results from competing models lead to similar decisions, then it is tempting to conclude that the decision to be made is robust even in the face of alternative theories. However, it can be the case that combining dichotomous model results may yield an answer that is very different from the answers that either model gives separately. For example, if two models predict nonoverlapping distributions of exposures to a given population (e.g., one model predicts potential doses from 0 to 100 µg/kg/day, and another predicts 500 to 600 µg/kg/day), then the average of the two may be a number which is not supported by either model (Finkel, 1989). Furthermore, it can be the case that all of the models are wrong for different reasons. Efforts to validate the models quantitatively, or to determine which model best represents the relationships in data sets, are the best confidence-building measures. Other confidence-building measures are more qualitative, and are discussed in Chap. 3.

One technique that helps analysts in evaluating models is influence diagrams. These diagrams are a graphical depiction of the relationship among quantities in a model that provides insight into a model structure. An example of an influence diagram is shown in Fig. 2.2. Influence diagrams can help turn a black box into a "grey box," in which the structure of a model may be made more transparent to a user. The structure is indicated by identification of the key inputs, intermediate variables, or submodels within the model, and the relationships between the components as denoted by arrows. An arrow typically implies that one component of the model is an input to another component of the model. The limitation of influence diagrams is that it is difficult to depict the details regarding specific functional relationships between variables or regarding the specific values of model inputs. However, one can choose appropriate symbols for nodes to distinguish among different types of input assumptions (e.g., deterministic, probabilistic, indices, etc.) and use different line thicknesses for the nodes to distinguish between variables and submodels. Software programs, such as Analytica® of Lumina Decision Systems, are available to develop models and influence diagrams simultaneously (Lumina, 1996). An advantage of an influence diagram is that it aids in communicating qualitative information regarding the structure of a model. Furthermore, it is possible to put together influence diagrams early in model development to help identify relationships among variables, to guide data

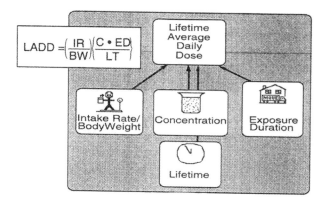

FIGURE 2.2 Influence diagram for a simple exposure model (Source: Frey, 1992).

collection and model building, and to facilitate communication among members of a project team.

2.2.1.2. Model Detail. Models are often simplified for purposes of tractability. For example, simplifying assumptions may be made to convert a complex nonlinear model to a simpler linear model. The simplified linear model may only be applicable to a much smaller portion of the problem domain addressed by the more complex version. Thus, the acceptable range of values for the inputs which will lead to valid model results will generally be more restricted than for the more complicated model. However, simplified models offer many benefits, particularly for policy applications. For example, they tend to be more transparent to the user, allowing more direct insight into system behavior. They may also be more amenable to analytical or numerical evaluation, enabling a fuller range of sensitivity and probabilistic analysis.

Uncertainty in the predictions of simplified models can sometimes be gleaned by comparison of their predictions to those of more detailed, inclusive models. In other cases, simple models are developed due to a lack of confidence or knowledge about what the actual model structure should be. In many simplified models, there is a heavy reliance on empirical factors (e.g., bioconcentration factors). In these cases, the simplified model is a signal that little is actually known or quantifiable about the phenomena being modeled. Uncertainty about these models may be only qualitatively understood.

2.2.1.3. Validation and Model Uncertainty. Key questions asked of modelers are "why should I believe your results?" or "is your model valid?" There are many levels of validity against which a model may be evaluated. Does the model do a reasonably good job of capturing the functional relationships among inputs to produce an output? For example, many models of the transformation of air pollutants in the atmosphere include a large set of nonlinear chemical kinetic equations in combination with a gridded three-dimensional representation of transport. An example is the Urban Airshed Model (UAM) for modeling of photochemical air pollution. A model based upon these chemical and physical mechanistic relationships can potentially provide a more effective representation of reality than a simplified or linearized model. For example, another air quality model, Empirical Kinetic Modeling Approach/ Ozone Isopleth Plotting Package (EKMA/OZIPP), can be set up to use a similar chemical reaction mechanism as the UAM, but treats the entire airshed at a much larger level of spatial aggregation than the UAM (e.g., Singh and Frey, 1998). Specifically, EKMA/OZIPP uses only one cell to represent the entire airshed, whereas UAM may use hundreds of cells to represent both horizontal and vertical variability in pollutant levels. Thus, the larger model in this case would appear to have greater validity to its structure or construction. However, we have not addressed the question of whether the model does a good job of predicting actual pollutant concentrations.

If limited data are available for estimating the inputs and parameters of a "large" model, then a simpler model could be more accurate at predicting observed ambient concentrations. For example, the average emissions over an entire airshed may be relatively well-known. In contrast, the variability in emissions from one 5 × 5 km grid cell may be more uncertain, due to lack of site-specific data. Thus, more assumptions would have to be made to develop the additional inputs required by the larger model. If these assumptions are not developed carefully, it is possible that the larger model may provide less precise or accurate predictions than the simpler model within the domain of model inputs for which the models have been validated. However, when making predictions outside of the domain of inputs for which the models have been validated, the simpler model may be more prone to errors. This may be due to an inadequate characterization of the physics and chemistry in the simpler model compared to the larger model. For example, the simpler model would not adequately account for three-dimensional air flow, or for the spatial variability in emissions, temperatures, and, hence, reaction rates, that leads to spatial variability in ambient pollutant concentrations. The validity of a model for making predictions is often what analysts and decision-makers care about most, but in many cases this is extremely difficult to evaluate.

Questions about model validation also must contend with potential differences between correlation and causality. Two quantities may vary together, but this may be due solely to the unknown influence of a third quantity. The correlation between the first two quantities is then said to be "spurious." Thus, a model developed based upon the measured data set can be completely invalid, although it may appear to be quite good. For example, a particular exposure may be correlated with an increase in a particular health risk. For this reason, an exposure assessment is developed that focuses on scenarios for a specific pollutant. However, the causal factor may be something else, such as another chemical that is overlooked in the scenario. Thus, the model based upon the wrong inference regarding the causal relationship between exposure and effect will be irrelevant, and reductions in exposure aimed at reducing the health effect derived from the model will be ineffective.

When empirical data pertaining to both the inputs and outputs of a model are available, the precision and accuracy of the model predictions compared to the empirical data set can be evaluated quantitatively. However, such an evaluation is based upon the assumption that the structure of the model adequately captures the causality of the system being modeled. For example, when comparing the predictions of a chemical kinetic model for air pollutant formation to experimental results, the differences between the model predictions and the actual observations may be due to measurement errors or to errors in the structure of the model. In the former case, errors in the input assumptions would be the cause of errors in model predictions. In the latter case, even if the input assumptions were adequate, the model may produce an erroneous prediction. Conversely, if a model produces an acceptable prediction, it could be because of: (a) reasonably correct input assumptions combined with a reasonably correct model structure; (b) compensating errors in the model input assumptions and a reasonably correct model structure; or (c) a combination of incorrect input assumptions combined with errors in the model structure, all of which compensate for each other. Of course, if the model structure is incorrect, then the model may be using irrelevant inputs or may not be using inputs that would be needed if the model formulation were improved. As an example of the latter, if a particular chemical reaction were missing from the chemical kinetic model, then the inputs associated with that missing reaction (e.g., activation energy, Arrhenius constant, stoichiometric coefficients) would also be missing. Thus, agreement between a model prediction and empirical values of the quantity being predicted is not a sole guarantee that the model is valid.

Uncertainties regarding models for which few data are available to test model predictions may require evaluation using expert judgment or may not be amenable to any quantitative characterization. Approaches to model validation are discussed in Chap. 3.

2.2.1.4. Extrapolation. A key source of uncertainty is extrapolation. Models which are validated for one portion of a parameter space may be completely inappropriate for making predictions in other regions of the parameter space. For example, an exposure model developed for characterizing chronic dose rates may be inappropriate for characterizing acute dose rates.

2.2.1.5. Resolution. In numerical models, a spatial and/or temporal grid size must be assumed. For example, in finite difference methods used to solve systems of ordinary differential equations, a time step must be specified. The solution is then calculated for each time step and is typically propagated forward through time. The accuracy of numerical methods is typically inversely proportional to duration of the time step. However, as the time step becomes very small, numerical round-off errors can become a dominant source of uncertainty in the solution (Hornbeck, 1975). Thus, there may be an optimal time step for a particular problem.

The purpose and desired accuracy of the model prediction may also be factors in selecting a time step. For example, in developing exposure estimates for acute health effects such as respiratory illness due to ozone, one faces a trade-off between accuracy associated with using short time intervals and the model run time, which would be shortened by using longer averaging periods. Thus, the selection of the step size involves a trade-off between computation time (hence, cost) and prediction accuracy. Standard techniques are often available to help select the appropriate step sizes for a particular target accuracy. This type of model uncertainty can be evaluated by comparing results based on different grid sizes. However, often the key cost associated with finer grid sizes is the need to obtain more input data, such as for smaller spatial segments or shorter time periods. Such data tend to have more uncertainty than when larger scale spatial or temporal averaging is employed. It may be necessary to grapple with the problems associated with data inputs for finer grid sizes because, for some types of problems, it may be necessary to consider the time and location history of pollutant concentrations to more properly quantify exposures related to a given health effect.

2.2.1.6. Model Boundaries. Any model has limited boundaries in terms of time, space, number of chemical species, temperature range, types of pathways, and so on. For example, an assessment of the formation of photochemical air pollution must be bounded by a geographic area and a volume of the atmosphere to be considered in the analysis (e.g., the volume of air over the selected geographic area up to some altitude above the mixing height), the beginning and ending times of the meteorological episode to be simulated, the ranges of wind speeds and directions, the primary and secondary pollutants to be considered in the chemical reaction mechanisms, and the variation in ambient temperatures as a function of time and space. The selection of a model boundary may be a type of simplification. For example, judgment is required regarding what altitudes of the atmosphere to include within the model. Within the boundary of the model and parameter space of the problem, the model may provide an accurate representation of the real-world phenomena of interest. However, other overlooked phenomena not included in the model may play a role in the real world scenario being modeled. For example, a photochemical air quality model that has as its boundary an urban airshed may fail to capture potentially important interactions associated with longer range transport of pollutants. Similarly, a photochemical air quality model may fail to capture local effects that result from acid deposition due to regional transport or other environmental problems that perhaps should receive equal attention.

The issue of model boundaries is also related to that of dependence among model inputs. If a model is narrow in scope, it may have inputs which are in fact dependent upon each other. For example, the emissions of volatile organic compounds in one grid cell of a photochemical air quality model may be dependent upon the emissions in adjacent or nearby grid cells because of the common influence of the local economy. In this example, the dependence is from some factor which is exogenous to the model, due to the selection of model boundaries which exclude endogenous consideration of the economy.

Econometric studies, for example, contend explicitly with the choice of independent (explanatory) and dependent variables when building an empirical model. Candidate explanatory variables may be rejected *a priori* because they are either irrelevant or not observable. In the former case, there is no effect on the validity of the model. However, in the latter case, the model validity may suffer. Explanatory variables may be evaluated with respect to hypotheses about how a system should behave. If the model output is expected to increase as one of its inputs increases, but based on a particular data set the opposite response is observed, then it is typical to search for artifacts of the data set. For example, a potential explanatory variable may not have varied over a significant enough range to affect the dependent variables, or the influence of the variable may have been mitigated by more important processes or effects. In the latter case, even a large variation in the candidate explanatory variable may lead to little variation in the dependent variable. Thus, even though a model may be constructed and validated based upon a particular data set, the selection of dependent variables can be such that the model would be invalid for making extrapolations to other values of the explanatory variables, due to the failure to properly account for additional explanatory variables.

2.2.1.7. Scenario Uncertainty. Prior to using a model, an analyst must develop (explicitly or implicitly) a scenario for the problem of interest. A scenario is a set of assumptions about the nature of the problem to be analyzed. For example, in many environmental problems, assumptions are made about the source of pollutant emissions (e.g., is it just the incinerator, or could there be other sources that also contribute?), the pathways of pollutant transport (air, plant uptake, groundwater, etc.), and the times and activities during which exposures occur (e.g., outdoor, indoor, residential, occupational, time of day, etc.). Scenarios may be constructed to represent an actual environmental problem, or they may be constructed hypothetically based on policy motivations. In the latter case, for example, a scenario may focus on a hypothetical "porch potato"—an individual who spends his or her entire lifetime at the point of maximum concentration of some pollutant. In the example given in Chap. 9, the scenario is carefully defined to provide results for a specific population. To the extent that the scenario fails to consider all factors affecting the key output variable (i.e., average daily dose averaged across a year), uncertainty will be introduced. Like the uncertainty associated with model boundaries, the uncertainty associated with the scenario can be addressed by imaginative thinking about all possible factors that come to bear in either the real-life or hypothetical problem.

2.2.2. Variability

Some quantities are inherently variable over time, space, or some population of individuals (broadly defined) rather than for any individual event or component. Variability is referred to in a number of ways by different people. Terms that you might encounter include stochastic uncertainty, aleatory uncertainty, or Type A uncertainty (Kaplan and Garrick, 1981; IAEA, 1989; Morgan and Henrion, 1990). Aleatory is of, or pertaining to, natural, accidental causes. Stochastic or aleatory uncertainty are types of variability that may not be explainable using mechanistic theory. For example, why some people are highly susceptible to health effects from a chemical in the environment, while others are not, may appear to be stochastic only because of lack of good theory or data sets to help explain the reason for the variation. In other cases, the use of frequency distributions to quantify differences in a population may be a matter of convenience. We may know why some people ingest more of a particular food than others, but there may be no need to develop a detailed model that explains such variation. In the case study of Chap. 9, variability must be characterized in all of the model inputs (e.g., consumption rates, PCB concentrations).

Variability exists, for example, in the body weights of adults. In some cases, there may be several distinct groups of individuals which are more nearly homogeneous than some overall population. In such cases, the observed variability may be well described by a mixture of frequency distributions for various subpopulations. Empirical data sets can be evaluated to determine whether they are comprised of a mixture of distributions (e.g., D'Agostino and Stephens, 1986; McLachlan and Basford, 1988). An example of the analysis of data obtained from a mixture distribution is given in Section 5.7.3. Insights regarding the components of the mixture can be useful in targeting highly exposed or highly susceptible subpopulations for further study.

2.2.3. Input Uncertainty

Uncertainty, like variability, goes by several names. Some common ones are epistemic uncertainty, lack-of-knowledge uncertainty, or subjective uncertainty. It is often stated that variability is a property of the system being studied, whereas uncertainty is a property of the analyst. Different analysts, with different states of knowledge or access to different data sets or measurement techniques, may have different levels of uncertainty regarding the predictions that they make (e.g., NCRP, 1996).

Morgan and Henrion (1990) have identified a number of different types of quantities used in models. These include:

- *Empirical Quantities*: Measurable, at least in principle (e.g., pollutant concentration).
- *Defined Constants*: Some quantities whose values are accepted by convention, such as Planck's constant or the speed of light, are actually empirical quantities subject to measurement error, albeit small. Other quantities are defined by convention and are for practical purposes not uncertain. These include, for example, the mathematical constant π (pi).
- *Decision variables*: These are parameters over which a decision-maker exercises control, such as the maximum acceptable emission rate for a given emission source. A decision-maker selects this value. Thus, it is not appropriate to treat this quantity probabilistically, unless one is trying to predict what some other decision-maker might do. Typically, the sensitivity of the result to different values of the decision variable(s) should be explored using sensitivity analysis.
- *Value parameters*: Represents the preferences or value judgments of a decision-maker. Examples include the discount rate and parameters of utility functions used in decision analysis.
- *Model domain parameters*: These are parameters that are associated with a model, but not directly with the phenomenon the model represents. For example, the spatial or temporal grid size is a model domain parameter introduced in numerical models.

Of these types of quantities, only empirical quantities are unambiguously subject to uncertainty. Examples of empirical quantities commonly found in exposure assessments include engineering, physical, chemical, biological, physiological, and behavioral variables. In the context of exposure assessments, such as the example in Chap. 9, empirical quantities typically include concentrations of pollutants, environmental transport (e.g., wind speed, dispersion), activity patterns, and intake rates. The other types of parameters represent quantities which often are more properly treated as point-estimates reflecting convention, the explicit preferences of a decision-maker, or a discrete quantity by its nature (e.g., grid size). Thus, we focus here on identifying sources of uncertainty in empirical quantities.

2.2.3.1. Random Error. This source of uncertainty is associated with imperfections in measurement techniques or with processes that are random or statistically independent of each other. Random error in the context of measurements is inversely related to *precision* (e.g., Bevington and Robinson, 1992). Precision refers to the agreement among repeated measurements of the same quantity. Typically, random error refers to the deviation of individual measurements from the average of the population of measurements. Random deviations, however, provide no insight into the presence of systematic error, which would affect all measurements similarly.

2.2.3.2. Systematic Error. The mean value of a measured quantity may not converge to the "true" mean value because of biases in measurements and procedures. Such biases may arise from imprecise calibration, faulty reading of meters, and inaccuracies in the assumptions used to infer the actual quantity of interest from the observed readings of other quantities ("surrogate" or "proxy" variables) (e.g., Bevington and Robinson, 1992; Mandel, 1969). The latter occurs when one assumes a simplified model for how a system behaves and then makes measurements accordingly. For example, if one assumes that a metal contained in a fuel entering a furnace exits either in the bottom ash or the fly ash, one may overlook the possibility of volatilization of the metals into the flue gas. Thus, measurements of just the bottom ash and fly ash may systematically underestimate the total discharge of the material. Other sources of systematic error include, for example, self-selection biases in voluntary responses to surveys. Random and systematic error are compared and illustrated in Section 3.5.1.

2.2.3.3. Inherent Randomness or Unpredictability. Some quantities may be irreducibly random even in principle, the most obvious example being simultaneous estimates of small particle location and velocity as affected by Heisenberg's Uncertainty Principle. However, this concept is often applied to quantities that are in principle measurable precisely but as a practical matter (due for example to limited technology or excessive cost) are not. For example, it is difficult to measure the partitioning of dioxin between vapor and particulate matter because the measurement instruments may influence the partitioning (Cullen, 1995).

2.2.3.4. Lack of Empirical Basis. This type of uncertainty cannot be treated statistically, because it requires predictions about something that has yet to be built, tested, or measured (Frey, 1992). This type of uncertainty can be represented using technically-based judgments about the range and likelihood of possible outcomes. These judgments may be based on a theoretical foundation or experience with analogous systems. In cases where data exist for analogous systems, it may be possible to fit probability distribution models to the existing data set. However, it is important to consider the effect of systematic differences between the measured quantity and the quantity of interest. This may motivate changing the mean or variance of the fitted distribution.

2.2.3.5. Dependence and Correlation. When there is more than one uncertain quantity, it may be possible that the uncertainties are statistically or functionally dependent. Failure to properly model the dependence between the quantities can lead to uncertainty in the result, in terms of improper prediction of the variance of output variables. In general, it is desirable to explicitly model the source of dependence between the quantities, with one being calculated as a function of the other. This topic is addressed more fully in later chapters, including Chap. 6 and Chap. 9.

2.2.3.6. Disagreement. Where there are limited data or alternative theoretical bases for modeling a system, experts may disagree on the interpretation of data or on their estimates regarding the range and likelihood of outcomes for empirical quantities. In cases of expert disagreement, it is usually best to explore separately the implications of the judgments of different experts to determine whether substantially different conclusions about the problem result. If the conclusions are not significantly affected, then the results are said to be robust to the disagreements among the experts. If this is not the case, then one has to more carefully evaluate the sources of disagreement between the experts.

In some cases, experts may not disagree about the body of knowledge. It may be the case that the experts use different mental models to interpret an agreed-upon body of knowledge. Thus, the differences in expert opinion may be reduced to clearly identified differences in inferences that the experts make from the data. In this situation, the experts, analyst, or decision-maker may have to make a judgment regarding which experts are more authoritative for the problem at hand. In cases with strong disagreements, it may be useful to consider what additional information would be required to resolve the disagreement.

2.3. COMPARISON OF VARIABILITY AND UNCERTAINTY

Both variability and uncertainty may be quantified using probability distributions. However, the interpretation of the distributions differs in the two cases. Kaplan and Garrick (1981) suggest that uncertainty regarding variability may be viewed in terms of probability regarding frequencies. The International Atomic Energy Agency (IAEA) (1989) interprets distributions for variable quantities as representing the relative frequency of values from a specified interval, and distributions for uncertain quantities as representing the degree of belief, or subjective probability, that a known value is within a specified interval. Of course, uncertainty can arise not due only to expert judgment regarding a quantity for which little data exists, but also due to random sampling error and measurement errors. Morgan and Henrion (1990) suggest that variability is described by frequency distributions, and that uncertainty in general, including sampling error, measurement error, and estimates based upon judgment, is described by probability distributions. Here, we adopt the terminology of Morgan and Henrion in using the term "frequency distribution" to represent variability, and the term "probability distribution" to represent uncertainty. Because the concepts of variability and uncertainty are distinct and can have different implications for decision-making, it may be necessary to consider them separately in an analysis.

In exposure assessments, a common source of variability are differences in characteristics between individuals (e.g., intake rates, activity patterns). However, there may also be uncertainty in the characteristics of specific individuals in the population, due to measurement error or other sources of uncertainty as described earlier. In these cases, there is a resulting uncertainty about the variability frequency distribution. For example, while a surveyed group of individuals may be known to have different levels of exposure to a certain pollutant, the distribution of a similar but unsurveyed group may be extrapolated from the surveyed group. Limitations of the survey combined with extrapolation of the survey results to a larger population may introduce uncertainty. Thus, the population distribution for exposures may be both variable and uncertain.

To complicate matters further, it is possible for variability to be interpreted as uncertainty under specific conditions. The probability of randomly selecting an individual with a given exposure is the same as the relative frequency of all individuals in the population subject to the given exposure. Similarly, stochastic variation in weather may lead to uncertainty in the ability to predict short term pollutant exposures for a given individual. Hence, in these cases variability represents an *a priori* probability distribution for the uncertainty in exposure faced by a randomly selected individual. However, except for such cases, there is always a distinction between variability and uncertainty. Understanding variability can guide the identification of significant subpopulations that merit more focused study. In contrast, knowing the uncertainty in the measurement of characteristics of interest for the population can aid in determining whether additional research or alternative measurement techniques are needed to reduce uncertainty. For problems where the distinctions between uncertainty and variability are important, it is desirable to separately characterize them.

In principle it is possible to represent a myriad of uncertain or variable factors probabilistically (Hodges, 1987), but propagating uncertainty and variability together in a Monte Carlo simulation of exposures may complicate the interpretation of the results. In general, uncertainty assessments in which input distributions comingle both variability and uncertainty yield a distribution of exposures applicable to a randomly selected individual. If the primary objective of analysis is an assessment of exposure to identifiable subpopulations, it may be useful to separate uncertainty and variability. One option is to generate exposure distributions which represent specific cases of the variable inputs. In this approach the variable inputs are assigned constant values relevant to specific identified exposure scenarios or receptors, while uncertain quantities are assigned distributions. The output distributions represent uncertainty in exposure, for a given averaging time, location, and/ or receptor, or subpopulation. Conversely, uncertain quantities may be assigned fixed values (whose relevance is discussed qualitatively) while variable quantities are assigned distributions. Finally, multidimensional probabilistic analyses may be performed if one wishes to produce uncertainty distributions for specific subpopulations, for example, those in the upper percentiles of exposure (Bogen and Spear, 1987; Frey, 1992; Hoffman and Hammonds, 1994; MacIntosh et al., 1994; Price et al., 1996a&b; Frey and Rhodes, 1996; Cohen et al., 1996; Brattin et al., 1996; Burmaster and Wilson, 1996, and others). Multidimensional probabilistic analyses are discussed in Section 7.6.

CHAPTER 3

APPROACHES TO MODEL UNCERTAINTY

What is a model? Mathematicians view models as a set of constraints restricting the possible joint values of several variables. At a more conceptual level, a model may be viewed as a hypothesis or system of belief about how a system works or responds to changes in its inputs. A model contains a structure which may be predicated on theoretical or empirical considerations. The purpose of the model is to represent as accurately *as necessary* a system of interest. The degree of accuracy needed depends on the intended model application. For example, models developed for screening purposes may not need to be accurate; they may only need to be conservative. A system may consist of several components, such as emissions release, transport, deposition, plant uptake, and ingestion. The structure of the model is embodied in the form of the equations used and in the selection of variables which are treated as model inputs versus internally (endogenously) calculated quantities (Kirchner, 1990). The structure of the model may impose limitations on its applicability, for reasons discussed in Section 2.2.1.1. Models are always a simplified version of the reality of the system being simulated. All models involve aggregation and exclusion. Only a model without any aggregation or exclusion could ever possibly be an exact description of the system being modeled.

When is a model necessary? One widely held view is that you do not need to use a model when you have actual data for the problem in which you are interested, as long as you are not interested in forecasting or decision-making. Thus, if you have personal exposure monitoring data and if no other information is required for the purposes of the assessment, then no model is needed. Even in this situation, however, models can be used to help in interpolating within a data set, by using a mathematical formulation of a theoretical or empirical construct shown to be consistent with existing data.

Often, however, models are used to extrapolate to parts of a problem domain in which data are not available. In such cases, modeling may be helpful in making a prediction beyond the range of the data upon which the model is based only if there is a plausible theoretical basis for the model. As a practical matter, extrapolation of a model can lead to highly erroneous results, since the assumed theoretical basis underlying the extrapolation may be incorrect. It is often convenient to blame the model when such errors are made. Some models may be incorrect if they are based upon erroneous science, improper mathematical formulation, improper solution methods, mistakes in coding, or other shortcomings. In these cases, then, the model should not be used. In other cases, models may be misapplied by analysts. For example, the Mobile5a emission factor model developed by the EPA is intended to predict area-wide average emission factors for highway vehicles. However, it is routinely misapplied to estimate emissions for individual segments of highway (e.g., Kini and Frey, 1997). In this example, the model is not necessarily incorrect, but it has been misused. All models are developed with some intended purpose. When the model is misapplied for a purpose other than for which it was intended, one should expect errors. It is always *your* responsibility to make sure you are using models for their intended purpose.

3.1. PURPOSE OF USING MODELS

Models are developed for different purposes, often with different decision-makers and/or different types of questions in mind. Examples of some of the different types of model applications include:

- *Screening Analysis*: A screening analysis is usually based on simple models that are not likely to underestimate exposures, or using values of model inputs that are believed to be highly conservative. The purpose of these analyses is usually to help a decision-maker identify key areas for focus in regulatory development, data gathering, or scientific research. These models are useful in drawing attention to those exposure pathways or scenarios of concern. When the use of such a model yields an exposure estimate less than some threshold of concern, model users should have a high degree of confidence that, even if a refined analysis were done, the actual exposure level would be less than the threshold of concern.
- *Research*: Research models are intended to improve understanding of the function and structure of real systems. They allow researchers to explore possible or plausible functional relationships and may involve many detailed mechanisms.
- *Assessments/Decision-Making*: Assessment models are intended to serve as tools for decision-making, such as for rule-making or regulatory compliance purposes.

Models developed or applied for any of the above purposes may be implemented in a variety of ways. These include:

- *Deterministic Analysis*: Many models are developed for the purpose of providing a point-estimate which may be intended to serve as an accurate and precise prediction of some quantity. The purpose of such analyses is to provide decision-makers with a best-estimate that can be used in comparison with other assessments to compare exposures. However, quantitative measures of the accuracy and precision of model predictions are usually not developed, because no information on model or input uncertainty is accounted for quantitatively. In the case of a screening analysis, the values of model inputs may be selected to lead to a conservative result for the model outputs. In the case of many regulatory models, a model is accepted by convention as the designated model to use for certain decision-making purposes. However, such models are not always accurate or precise, and in many cases their accuracy and precision is unknown.

- *Sensitivity Analysis*: This is used to measure the potential importance of model inputs as contributors to variation in model outputs. Such an analysis can provide insight into whether (or how) a real world system is sensitive to perturbations of some of its components or processes, assuming that such relationships are adequately represented in the model. As discussed in Chap. 8, conventional sensitivity analyses have several limitations, such as difficulty in evaluating the effect of simultaneous changes in large numbers of model inputs. However, sensitivity analysis can be a useful first step in identifying model inputs that should be assigned distributions to represent variability or uncertainty in a probabilistic analysis.

- *Probabilistic Analysis*: Assuming that the model formulation is adequate, probabilistic analysis can be used to propagate uncertainties in model inputs to estimate uncertainties in model outputs. Unlike sensitivity analysis, probabilistic analysis yields quantitative insight into both the possible range and the relative likelihood of values for model outputs. One purpose of probabilistic analysis is to characterize variability and uncertainty in model outputs. This application helps decision-makers understand the range of exposures that are faced by different members of a potentially diverse exposed population. Another purpose is to identify key sources of uncertainty and variability that can be the focus of future data collection, research, or model development activity.

Many models developed for policy applications face the problem of *trans-science* as defined by Alvin Weinberg (1972). A trans-scientific question is one that can be asked in scientific terms but cannot, as a practical matter, be answered using the traditional tools of science. Questions about future exposures tend to fall in this category. For example, to answer the policy questions about global warming (e.g., "In what direction and by how much will global average temperature change over the next 50 years as a result of a particular scenario for greenhouse gas emissions?") would require conducting the actual experiment and waiting to get the answer. But this strategy would not be able to provide an answer to today's questions about how to avert such a future. Alternatively, if we had some extra earths in our same orbit and the leisure of a long period of investigation, we could conduct the experiment on those planets instead of ours (if ethical issues are set aside). However, to develop approximate answers to the questions requires taking indirect approaches, which involve model development, hypothesis testing of components of a much larger systems model, and so on. It is not possible to validate such models, in the strictest sense, because no empirical observations are possible at this time. Because policy questions often arise before the relevant science is well in hand, it is not uncommon that validation is an elusive goal, and that uncertainties about the formulation of the model itself may be more significant than uncertainties in the inputs to the model. The advantage of model development is that it allows us to construct what seem to be credible statements about the way a system works and to explore many "what-if" scenarios to see how things might turn out.

3.2. MODEL COMPLEXITY

A modeler typically has a choice regarding the level of complexity for the model structure. The appropriate level of complexity of a model depends on its intended applications. The complexity is characterized by: (i) the number of compartments, pathways, or states represented in the model; (ii) the number of inputs; and (iii) the functional form of equations. For example, equations may be linear time-independent or nonlinear and time-dependent. The former may be typical of exposure models developed for chemicals which result in chronic health effects, while the latter are developed for situations involving acute health effects or for situations in which feedback mechanisms or kinetics are important. Adding more mechanisms to a model typically leads to an increase in the number of inputs.

Complexity and size are two different issues. One may have a very large, but mathematically simple model, in which there are many inputs but the calculations are linear. On the other hand, one may have small, complex models, which do not require a great deal of computer code but contain intensive interactions among the components. Complex systems are often hierarchies, which can be described in terms of the "span" of each level in the hierarchy, and the number of levels. A simple system may have many repetitive components at only one level in a hierarchy (e.g., endless segments of bookshelves in the Library of Congress) and, hence, can be very large without being complex. Complex systems tend to be "made up of a large number of parts that have many interactions," and in particular tend to be comprised of hierarchies of subsystems. In many cases, complex systems are "nearly decomposible" into subsystems that have only weak interactions with each other, compared to more intense internal interactions within the subsystem (Simon, 1996). This latter property can aid in comprehending and analyzing a system.

Simple models may have a limited scope of applicability, because they do not account for interaction with other significant components of a larger system, but within that domain they are highly reliable and easy to use. For example, models used for screening studies are often of low complexity but rarely yield results that would be exceeded by the actual system being modeled. A specific example of this is the SCREEN3 air quality model, which is intended for use in preliminary assessments of ambient air quality impacts from air pollutant emission sources. This model addresses short-range transport under a limited set of scenarios. Therefore, it does not, for example, consider effects of pollutant transformation or allow for a particularly detailed treatment of topography. SCREEN3 is intended to produce conservative estimates, which should be somewhat higher than actual pollutant concentrations. The purpose of screening techniques is to eliminate the need for additional detailed modeling for situations that clearly are not of sufficient concern to warrant further action (USEPA, 1996c; USEPA, 1995b). An advantage of simpler models is that they tend to have fewer inputs, which reduces the opportunities for data entry or coding errors and reduces the time and effort required for the screening analysis. Such models, however, are often intended to be biased and cannot be used for predicting actual outcomes. In other cases, simple models may work better than more complicated ones if they use inputs that can be more easily quantified.

3.2.1. Uncertainty Propagation Characteristics of Models

Complex models are often intended to represent the best science which can be brought to bear on a particular problem. As such, one of the goals of using such models is to give quantitative expression to relevant theory and to produce predictions that are accurate. However, models that are more accurate in a structural sense typically have a much larger number of inputs than the simpler models. In going from simple to more complex models, uncertainty due to the structure of the model may be reduced; however, uncertainty due to the larger number of inputs, each of which may have its own estimation error, tends to increase. Ideally, the overall effect would be a reduction in uncertainty of the model predictions.

Complex models may be linear or nonlinear. Some nonlinear models may tend to have a magnifying effect on uncertainties in their inputs compared to linearized formulations. In other cases, nonlinear models may lead to results that are relatively robust even when the model inputs are highly uncertain. For example, many natural systems are homeostatic, meaning that they have a tendency to remain in a stable equilibrium, as long as the perturbations on those systems are within reasonable bounds. The variance in predictions obtained from some complex nonlinear models that amplify uncertainties may tend to be higher than for a simple model, due to the larger number of uncertain or variable inputs and the error propagation characteristics of the model structure. The result in this case could be a trade-off between model complexity and accuracy versus the variance of model predictions. In other words, in this situation the model may be more accurate but less precise.

As an example, the development of pharmacokinetic models is a means for developing and codifying insights regarding mechanisms by which contaminants reach target organs. Processes for the uptake of chemicals from the air and soil by vegetation, and contaminant migration in soils, have also been the subject of the development of complex models. However, the inputs required for such models are often more than what can be supplied based upon available data. Typically, as the level of detail of such models increases, the number of parameters for which data are available does not necessarily increase proportionately. Thus, additional judgments may be required to supply values for model inputs. A simple model based upon available data may do a reasonable job of interpolating within the existing data sets, without the overhead required to run or interpret results from a more complicated model. Especially if simple models are available which have been validated, their use should be seriously considered over the use of less well-validated but more complicated models. The more complicated models may in some cases serve well in helping to generate insights as part of research, while the simpler models may be more useful for routine assessment applications.

Ideally, the development of increasingly complex models should be accompanied by a reduction in prediction error variance compared to less complex models with fewer inputs. For example, consider a least-squares regression analysis. Typically, one would add additional independent variables to the model only if the standard error of the estimate of the model predictions, or the coefficient of variation of the model, decreases. This indicates that some portion of the variance in the model output can be explained by including the additional independent variable.

The mathematical form of a model plays a key role in determining uncertainty in predictions based on uncertainty in inputs. Consider a simple comparison of uncertainty in a model output, Y, due to the same uncertainty in the model input, X, but for different model structures. Let X be a normally distributed random variable with a mean of one and a coefficient of variation of 0.3. The resulting uncertainty in Y is shown graphically in Fig. 3.1 for three alternative model structures:

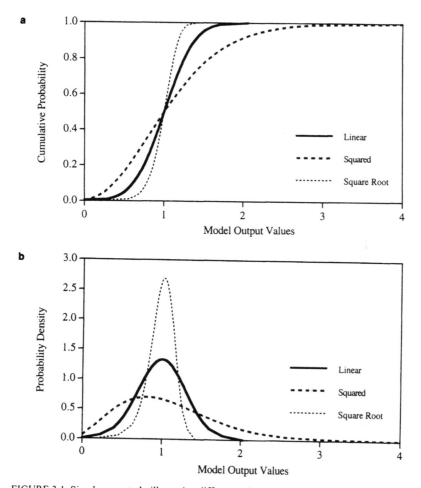

FIGURE 3.1 Simple case study illustrating differences in error propagation features of alternative model formulations. (a) Model results shown as cumulative distribution functions. (b) Model results shown as probability density functions.

(1) Linear model, $Y = aX$;
(2) Squared model, $Y = aX^2$; and
(3) Square Root model, $Y = aX^{1/2}$.

The results are shown as both cumulative distribution functions and as probability density functions. In all three cases, $a = 1$. Thus, the probabilistic result for the linear model has the same numerical values as the input distribution for X. As illustrated in the figure, compared to the linear model, the squared model leads to a much wider range of variation in the output and significant positive skewness in the distribution for the output. The skewness (see Section 5.1.3 for definition and discussion) is most evident in Fig. 3.1(b), where the result for the squared model has a long tail to the right. In contrast, the square root model leads to a much narrower range of uncertainty for the model output and a negatively skewed distribution. The negative skewness is most evident in Fig. 3.1(b), in which the distribution for the square root model has a slightly longer tail to the left than to the right. This simple example clearly illustrates that the selection of model form can affect error propagation characteristics.

Of course, in addition to the model structure, the magnitude of the input and its role in the system of equations must also be considered. Some models, such as ecological models, may be structured such that they tend to amplify errors in the inputs. For example, 4% error in model inputs led to 12 to 40% error on model outputs in one ecological risk assessment study (O'Neill et al., 1980). The error propagation properties of a nonlinear model can lead to a lack of precision, which may turn out to be more important than the lack of accuracy which may be associated with an erroneous mathematical form (Gardner et al., 1980). Thus, more complex or nonlinear models can lead to higher variance in model predictions than a simplified linear version. This outcome is not true, however, of all complex or nonlinear models. The error propagation features of a specific model should be studied and considered in each case.

3.2.2. When Is a Complex Model Needed?

In some cases, models evolve to increasing states of complexity due to the competing demands made upon them. For example, regulatory models may tend to be complex because they are used for a variety of different applications. Consequently, they must include a variety of mechanisms that allow the models to be adapted from one application to another. Such models often end up being so data intensive or user unfriendly that they become cumbersome to use. Thus, they serve no one well (Morgan and Henrion, 1990).

Complex models are often justified in situations where the details of the system being modeled are well-understood and in which inclusion of the details is necessary to get the right answer (Morgan and Henrion, 1990, p. 301). For example, to characterize urban air quality, it is necessary to use large, complex photochemistry models such as UAM-V. Large models may also be justified in the context of research to explore alternative mechanistic assumptions and for codifying the current state-of-knowledge about a particular problem.

Many experienced modelers offer the following advice for those developing models for policy analysis applications: Make your model (or theory) as simple as possible, but no simpler. In other words, include whatever you need to best represent the relevant science, but keep the model simple enough that it is easy to use.

3.2.3. Simplifying a Complex Model

One may simplify a complex model by developing a response surface representation of it. A response surface is a simplified representation of the relationship of a selected number of model outputs to a selected number of model inputs, with all other model inputs held at fixed values. The development of a response surface typically involves extensive sensitivity analysis of the complex model to generate a calibration data set, developing alternative simplified model equations, fitting the parameters of the equations using the calibration data set and methods such as regression analysis, and evaluating the goodness-of-fit of the simplified model compared to the complex model. For example, Frey and Bharvirkar (1998) have developed a response surface model of a detailed chemical process simulation model of advanced coal-based power generation systems. The simulation model required approximately a day of CPU time on a dedicated workstation in order to generate 100 sets of model inputs and outputs for use in developing the response surface model. The values for a total of 12 simulation model inputs and 60 simulation model outputs were collected for the calibration data set, and the simulation model was run with various combinations of values for the 12 inputs. A total of 60 equations, one for each simulation model input, were fit to the data set using linear regression analysis. Each equation provides a prediction for a model output as a function of up to 12 of the selected model inputs. The response surface model takes only seconds to execute on a desktop computer. The predictions of the response surface model are accurate to within plus or minus one percent in most cases, compared to the simulation model. The precision of the predictions of the response surface model are within plus or minus a few percent. Thus, at the expense of very little additional prediction error, an overall benefit of substantially faster run times was achieved.

A disadvantage of response surface models is that their range of applicability is typically less than that of the complex model, because not all valid combinations of input assumptions may be considered when developing the calibration dataset. For example, in the previous example, only 12 user-specified inputs were allowed to vary within defined ranges, while hundreds of other input assumptions were held constant. Thus, the response surface model developed based upon the data set is not valid for any extrapolations beyond the range of input values used as input to the complex model. However, it is typically the case that only a handful of variables, even in complex models, drive the answer. Thus, such simplifications may retain much of the predictive power of the complex model, while offering the benefits of transparency and ease of use.

In the context of probabilistic analysis, response surface models are more amenable to input uncertainty analysis. For example, Constantinou et al. (1992) developed simplified response surface models based on sensitivity analysis of complex fate and transport, deposition, exposure and dose, and health risk models. For some purposes, it is reasonable to simplify an atmospheric transport model to a function of the chemical emission rate and a multiplicative constant. Constantinou et al. estimated such a constant based upon the particular meteorological data, topographical information, and source characteristics needed for a specific assessment. The simplified model facilitates the analysis of uncertainty in chemical concentration in the air due to uncertainty in emission rates, assuming that all other parameters are fixed. More complex response surfaces can be constructed as appropriate.

Probabilistic modeling techniques can be employed to help develop a response surface model. Typically, you would want to run the complex model many times to generate a paired data set of model input and output values. Then these values can be used to construct a graphical response surface, or you can use multivariate regression analysis techniques to fit a simple model to the generated data. If one treats the complex model as having made completely accurate and precise predictions, then the error in predictions made by the regression model may be evaluated with respect to the predictions of the complex model. Thus, the standard error of the estimate for the regression model is a measure of the uncertainty in model predictions due to the use of a simplified function form (e.g., Frey, 1993). Ideally, the response surface model will be accurate, but may have less precision than the detailed model. In reality, it is likely that there will also be biases in the predictions of the simplified model, at least for some parts of the input domain.

Another approach to model simplification is to take a mathematical approach, in which one develops simplified functional forms of the equations of the complete model. For example, one might linearize model behavior over some portion of its domain. The simplified linear model can then be used more easily than the parent model when applicable. The development of simplified reduced form models has been done for the purposes of integrated assessment of acid deposition. For example, Small et al. (1995) developed a simplified "meta-model" of a large model known as Model of Acidification of Groundwater in Catchments (MAGIC). Rubin et al. (1992) developed an integrated assessment model that contains several meta-models for major components of the acid deposition process. More recently, a team of investigators have developed the Tracking and Analysis Framework (TAF), which contains meta-models of major components required to assess the impacts of acid deposition (Henrion et al., 1995). Integrated assessment of the impacts of acid deposition requires quantification of current and future national emissions of sulfur dioxide and nitrogen oxides, physical and chemical mechanisms affecting pollutant atmospheric transport and transformations, deposition process, and effects processes at various receptors such as lakes, soils, humans, materials, and others. For each of these, large and detailed models are either available or could be constructed. However, it is necessary to develop reasonable but simplified representations if all of these are to be included in one modeling framework.

Model simplification can also be accomplished by taking different approaches to model development. For example, McKone (1996) compared two models for estimating the inventory and flux of contaminants in soils. One model was based upon exact solution of the set of dispersion and advection differential equations that describe contaminant fate in the soil. A second model was based upon a simpler mass balance approach with less spatial resolution. It was demonstrated that the two models represented the variability of the movement of trichloroethylene (TCE) in the soil with comparable reliability. Thus, an analyst may wish to choose the model that is easiest to work with or requires the least amount of input data.

3.3. SETTING UP THE RIGHT PROBLEM

Model uncertainty is usually connoted with a lack of confidence that a mathematical model is a "correct" representation of a particular problem (Hammonds *et al.*, 1994). However, there are really two dimensions to this issue: (1) differences between the model boundaries (or the model domain) and the boundaries for the scenario of interest; and (2) the adequacy of the mathematical model in representing the domain for which it was developed. We can illustrate these distinctions with a drawing developed from similar presentation in Kirchner (1990).

Using a Venn diagram, Fig. 3.2 illustrates the difference between the domain of a system being studied and the domain of the model developed to represent the system. If you think of the mathematical model as an approximate rep-

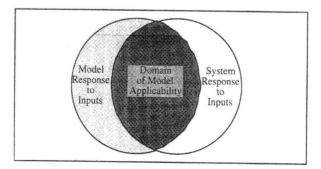

FIGURE 3.2 Venn diagram representing the difference between model and system response to the same set of inputs (Kirchner, 1990).

resentation of the system, then you can imagine that both the system and the model would be responsive to the same set of inputs. However, because the model is only an approximation of the system, sometimes its response to the inputs will be different than the system's response. The model is a good representation of the system only when its response is like that of the system. This situation is shown as a region of overlap between model response and system response in the Venn diagram.

If we consider that the model was developed to represent a particular system, we realize we can introduce yet another source of error by applying the model to simulate a different system. Thus, a model which is demonstrated to work well for a particular system may fail completely when applied outside its intended domain. For example, a simple Gaussian-based model for air pollutant dispersion may work well for a stationary source with a homogeneous and turbulent atmosphere. However, without empirically-based corrections to the parameters of the model, the model cannot be applied to atmospheric situations of varying stability (Seinfeld, 1986).

3.4. MODEL VERIFICATION

There are several major steps to model verification. The first is to make sure that the mathematical model is correct. This involves ensuring that the mathematical structure of the model is free of error. The second step is typically to make sure that the computerized version of the mathematical model is correct, so that it reproduces the desired mathematical structure of the model and yields the proper results for a given set of inputs. The third major step is to make sure that the values of the inputs are correctly specified. A discrepancy between the predictions of a computer program and the expected result for the model output may be due to errors in the structure of the mathematical model, the coding of the model into the computer, or the specification of model inputs. If a discrepancy is not observed between model predictions and expected results, this does not mean there are no errors in the model. It is possible that there are compensating effects due to multiple errors in the mathematical structure, computer code, or inputs. Thus, it is important to verify components of the model, and not just the entire model.

One of the more labor-intensive steps in model development can be the verification of computer code. Verification of code typically involves one or all of the following types of activities (Rish, 1988):

- Comparison of computer code calculations to hand calculations.
- Comparison of computer code calculations with an alternative calculation scheme.

- Comparison of computer code calculations with another computer code which has been verified.
- Conducting a detailed and independent review of the computer code.

For simple models, verification via hand calculations may be feasible and will assist in revealing not just the presence of an error in model predictions, but also the source of the error.

In many cases, there may be more than one way to arrive at a model prediction. For example, there may be an analytical solution and several alternative numerical methods for solving the same problem. Comparison of computer code results with results obtained from other methods can detect the presence of an error. If a numerical model is coded, it is a good idea to set up at least one simplified problem that can be solved analytically. The analytical results and computer code results should, of course, agree, or an error exists in the model implementation. The comparison of a new computer code to one that has been previously verified is called "benchmarking".

In some cases, verification may be more in the form of an expert judgment than a quantitative analysis. Such verifications may be appropriate when you are only interested in the qualitative behavior of the model. Thus, you would look to make sure that the output changes in some expected way as a function of changes in inputs. This can be done using sensitivity analysis.

Regression-based sensitivity analysis used in combination with probabilistic modeling techniques can be a powerful model verification tool. This technique, described in Chap. 8, enables the user to identify the probabilistic inputs which are the most important contributors to variation in uncertain model outputs. Thus, this approach provides an indication of the most sensitive variables in the model. The set of sensitive variables obtained from such regression analyses should be compared to expectations regarding key uncertainties. If the results are in line with prior beliefs about the behavior of the model, then the model user may develop increased confidence in the model performance. If the results are counterintuitive, then the model user should investigate whether the results are merely unexpected but reasonable or whether they reflect an error in the mathematical structure of the model or the implementation as computer code.

Computer aided software engineering (CASE) tools are emerging which help automate the process of model verification. For example, tools exist for performing dimensional analysis on a model to determine if the desired units for model outputs can in fact be obtained based on specified units for the model inputs and the structure of the model. A lack of dimensional closure implies an error in the model formulation (Kirchner, 1990; Cmelik and Gehani, 1988; Hilfinger, 1988).

Of course, there may be no substitute for careful line-by-line verification of a computer code. In one case, for example, a large assessment model survived comparisons with hand calculations and with other models. Afterwards, it was discovered that a coding error reversed wind directions. This error was found a year later by someone who studied a spreadsheet listing results for different geographic locations. In the meantime, the model had been used for analyses and assessments. This scenario is the nightmare of many modelers.

3.5. MODEL VALIDATION

Model validation is a process of hypothesis testing to determine whether a model can be rejected as false. The type of validation that is needed depends on the purpose of the model. Some models are developed only for defining functional relationships between quantities and the overall behavior of a system. In such cases, the model users may be more interested in evaluating relative changes in model predictions as a function of changes in input assumptions than in making accurate predictions. For example, ecological system models may be developed to evaluate the stability of predator–prey relationships. Such evaluations may not require precise, site-specific input values (e.g., Gardner *et al.*, 1980). In these cases, the models are not used to make predictions so much as they are used to help generate insights into the qualitative features of the behavior of a system. Other models may be developed for applications such as the development of sampling schemes (e.g., Kirchner, 1990). The actual data used in such models does not have to represent any specific system, but should fall within a reasonable range. It is only when a model will be used for making a prediction that it should be subject to a full validation process. Validation is an analysis that can reveal conditions under which a model fails to perform adequately. While the process of validation cannot prove that a model is a truthful representation of a system, it can lead to falsification, or invalidation, of a model.

3.5.1. Validating with Data: The Elusive Ideal

Model validation is the comparison of model results to observations from the system being modeled. The analysis associated with validation is also useful for estimating uncertainty due to the model. Given a set of inputs for both a model and a system, one can compare the model response and the system response. The difference between the two can be used to characterize model accuracy (bias) and precision (random error). Accuracy and precision are illustrated in Fig. 3.3. A model that is accurate will generate values that, on average, correspond to the "true" value of the quantity for which the predictions are developed. When there is a discrepancy between the true value and the average result obtained from the model, then a bias exists. Precision refers to the agreement among repeated predictions. Fig. 3.3 (a) illustrates one case in which a model is accurate, with no tendency for the predictions as a group to be too high or too low, but imprecise.

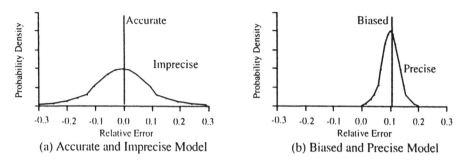

(a) Accurate and Imprecise Model   (b) Biased and Precise Model

FIGURE 3.3 Possible characterizations of the precision and accuracy of the predictions from a model with respect to known values for the model outputs.

In contrast, Fig. 3.3 (b) illustrates a situation in which a model is biased. Bias is also referred to as constant error or systematic error. In this example, the biased model produces precise results. However, nearly all of the predictions are too high compared to the true value. Ideally, a model will be both accurate and precise. However, if a bias is known, it is possible to adjust for the bias. The process of validating a model by comparing to known values for model inputs and model outputs can also yield estimates of model precision and accuracy.

When there are competing models, which may represent alternative hypotheses about how the system of interest works, the multiple models may be evaluated with respect to an independent data set that was not used in calibrating any of the models. The data can be used to "pick" the model that best represents the data. For example, the model with the best accuracy might be preferred over other models that may be more precise but less accurate.

As a practical matter, validation can be very difficult. For large models, validation may be possible only for some components of the model, or for only some parts of the problem domain. In situations where spatial and temporal variability is important, data are usually available only for relatively short time periods or at a few locations. For example, in radiation exposure assessments, model validation may be nearly impossible due to extremely low levels of radionuclide concentrations in the environment or because the time periods considered by the model are prohibitively long (Hoffman and Miller, 1983).

Hoffman and Thiessen (1996) report on a model validation exercise involving independent groups of modelers in a "blind testing" procedure. Each group was asked to make predictions regarding interindividual variability in ^{137}Cs concentrations in humans due to exposures that resulted from the Chernobyl nuclear reactor accident. The modelers' predictions were compared to measured values, which were withheld during the study. There were several interesting findings from the study. One is that most of the participants in the exercise did not try to characterize interindividual variability, but instead reported values intended to represent the arithmetic mean of the population. For those groups that did characterize both variability and uncertainty, there was a tendency to overestimate variability and underestimate uncertainty. Some of the groups that obtained what appeared to be accurate results did so due to the presence of compensatory errors. Hoffman and Thiessen caution that model performance in the exercise was for a specific assessment as performed by specific groups of modelers. They point out that a model may perform differently when used by a different group of assessors and for different assessment questions.

3.5.2. Partial Validation: When You Have Some Data

Situations in which models are tested against data only for some parts of the problem domain are known as partial validation. Similarly, validations based on comparisons of predicted behavior to the measured behavior of analogous systems are also only partial. It may be possible to make theoretical arguments for asserting that partial validation implies overall validity of the model (Rish, 1988). For example, photochemical grid models contain several components, including emissions, meteorology, and chemistry. The models employ simplifications to enable solution of physical and chemical equations within reasonable limits of computation time. It is generally not possible to fully validate this type of model. However, it may be possible to validate sections of the model for idealized situations (e.g., measurements from smog chambers for the chemical kinetics portions of the models).

3.5.3. What If There Are No Data?

In the absence of model validation, a model constitutes an untested hypothesis (Hoffman and Miller, 1983).

In situations where model validation is a practical impossibility, the next best alternatives include: screening procedures to identify potentially important model inputs and exposure pathways; sensitivity analysis to identify important groups of model parameters; probabilistic analysis to determine the effect of uncertainty in inputs on model predictions, and comparison among predictions of different models. These techniques help focus resources on the parts of the model that matter the most in an analysis, and they provide some insight into the variance in model outputs that is due to uncertainty and variability in model inputs. However, a shortcoming of this approach is that it does not, and cannot, provide insight into the accuracy of model predictions. Thus, while one can evaluate the implications of propagating variance through a model, without some type of external datum, such as that provided by relevant field measurements, one cannot characterize systematic errors. When there are no validation data, it is important to acknowledge this shortcoming in reporting results.

3.5.4. Can You Validate by Comparing Models?

Unless the model which is to be used as a benchmark has itself been validated, the comparison of models can be highly misleading. Even if the models agree, they may be subject to the same biases. Thus, both of their predictions could be systematically different from the actual values in the system being modeled. For example, in some cases in which multiple models exist to describe the same problem, they may have some common lineage (i.e. they may be based on similar theoretical foundations or mathematical equations). Thus, they would be expected to give similar results and yet provide no insight into whether either model is appropriate or biased.

It is also possible that two models may give similar numerical results, but for different reasons. For example, one model may deal with long term variability while the other deals with short term variability. If both sets of reasons should be considered as factors, then the implication is that one should develop a new model that combines the best features of the two original models. It is likely the results from the new model would be different than those of the original models.

Comparisons of the predictions of different models may be helpful in identifying errors in the structure of the mathematical models, the computer code, or the model inputs.

3.6. EXTRAPOLATION AND UNCERTAINTY

Models are developed on the assumption that their inputs will be contained within a range of values. Furthermore, field measurements or other types of observations required for validation purposes may exist only for a subset of possible model domain applications. Therefore, a model is only strictly applicable within the specific set of input and output values associated with the data used for validation. Extrapolation refers to situations in which the model is used for combinations of input values outside the range of data used for calibration or validation, or situations involving phenomena that cannot be described by the model due to limitations of its structure. For example, Gaussian-plume-based air dispersion models are typically considered applicable only for distances up to about 20 km or 50 km from the emission source. The models should not be used to characterize long-range transport over hundreds of kilometers or more.

Extrapolation is explicit when the selected values for model inputs are outside the range of values used to validate the model. However, there is also the possibility of hidden extrapolation, which occurs when the model is used for combinations of input values for which validation has not been done, even though all of the input values fall within the range of values which have been tested. Thus, range checks on each input variable will not guarantee the avoidance of hidden extrapolation.

Empirical models, such as ones developed from regression analysis on a data set, should not be extrapolated without good justification. In some cases, regression models are developed using arbitrary functions that may reasonably approximate the relationship between data in a specific data set. However, there may be no theoretical basis for the functional relationship that was selected. Thus, there would be no reason to expect that, for values of the independent (predictive) variables outside those of the original data set, the model would yield valid estimates for the dependent variable. An exception to this could be for a case in which the selection of functional form for the model was based upon defensible theoretical considerations. The regression analysis would be used to estimate the parameters of the theoretically-based model. A theoretically-based model, for which there is a high degree of confidence in the understanding of underlying physical, chemical, and biological processes and mechanisms, may perform reasonably well when extrapolated.

Models developed for deterministic analyses using point-estimates for all inputs may perform poorly when ranges are placed on all inputs. This may be particularly true of dynamic models. If a model was not originally developed with an intended purpose of probabilistic analysis, then the model user should proceed cautiously. The use of distributions rather than point-estimates could lead to extrapolations beyond the range of applicability of the model. This is one reason why it is important to consider probabilistic analysis at the beginning of a model development effort. Another reason is that in the process of doing a probabilistic analysis, it is possible to sample a joint set of model inputs which may correspond to some type of singularity point in the model, leading to division by zero or to nonsensical results. Typically, this can be traced to a simplifying assumption used in model development which may be violated in the probabilistic analysis. This type of problem is of special concern when adopting someone else's computer code for use in a probabilistic analysis.

3.7. CAN INPUT UNCERTAINTIES REFLECT MODEL UNCERTAINTY?

A model parameter is a quantity whose value is typically selected based upon a calibration or curve fitting exercise. For example, the parameters of a probability distribution may be selected using statistical estimators or probability plots. Similarly, the parameters of a model, such as an exponent or pre-exponential term in a nonlinear equation, can be estimated using regression analysis or via other comparisons with data. Implicit in such parameter estimation activities is the assumption that the structural form of the model is a good representation of the system which generated the empirically-observed data. If the structural form of the model is adequate, then uncertainty in the parameter values will generally be related to uncertainty in measurements of model inputs and outputs. In some cases, this may be the most challenging aspect of modeling. For example, in nuclear safety risk assessments, the scale on which data can be measured (e.g., over tens of years) may be very different than the scale upon which model predictions must be defined (e.g., thousands of years). However, if the structural form is not correct, then there will be both bias and variance in the parameter estimates. These uncertainties will be due to a combination of the effect of uncertainty in model inputs and outputs, as well as to uncertainty introduced by the model structure. For example, if you have a data set on body weight and intake rate, and assume that there is a linear relationship between the two, you might use linear regression analysis to determine the values of the model parameters. These parameters are the slope and intercept. Yet, if the underlying relationship between the two variables is really nonlinear, then additional uncertainty will be introduced into the estimates of the model parameters because of the failure of the linear model to capture this nonlinear behavior. The uncertainty in the parameters can reflect a portion of the uncertainty associated with incorrect model structure, but only a portion. Careful consideration of the implications of alternative model structures is thus recommended.

3.8. WHAT DO YOU DO WHEN MODELS DISAGREE?

Morgan and Henrion (1990) argue that every model is definitely false. The basis for this claim is that all models are incomplete descriptions of a real-world system. If one accepts that all models are false, then it is illogical to make statements regarding one model being more "probable" than another in representing reality, since all models are abstractions of reality. Thus, Morgan and Henrion recommend two approaches to dealing with model uncertainty when there is disagreement between models: (1) develop a meta-model of which the alternative available models are special cases; and (2) deal with each model separately and present all sets of results to the decision-maker. In the former case, model uncertainty could be converted to uncertainty in an input. For example, if there is disagreement over whether a system has a linear or exponential behavior, this type of model uncertainty could be expressed as a range of values for the exponent in question. In the latter case, it would be left to the decision-maker to make a choice regarding which model is plausible or whether to combine model results using some type of weighting scheme (e.g., Evans et al., 1994a).

Others disagree with the view espoused by Morgan and Henrion. Some argue, for example, that although the statement that all models are false is true in an absolute sense, we are often more interested in the ability of models to make predictions within specified criteria for precision and accuracy. Thus, a model that is able to meet objectives for accuracy based upon measurable criteria is a useful model. However, in many cases, objective measures of model accuracy may not be available. In such cases, an analyst may not have an empirical basis for selecting among competing models.

Another approach to dealing with alternative model formulations and structures is to ask experts to make judgments about the relative plausibility of each alternative. One can then employ a probability tree to calculate the relative plausibility of alternative models or cascades of models (Evans *et al.*, 1994a).

If significant sources of disagreement exist as embodied in different models, then it is always important to make this information explicit and communicate it to the decision-maker. Sensitivity analysis of weighting schemes can be used to provide some qualitative indication of how much preferences among the competing models would have to change in order for a decision to be affected.

3.9. EXAMPLES OF MODEL UNCERTAINTY ISSUES

In this section, we briefly describe an example in which model uncertainty is apparent. The example is based upon the Mobile5a emission factor model, which was developed by the United States Environmental Protection Agency.

Mobile5a is a model that is widely used to develop highway vehicle emission factors as part of State Implementation Plans (SIPs) and for other regulatory purposes. To use Mobile5a, a user must specify values for a large number of model inputs. Examples of the model inputs include, but are not limited to: average vehicle speed, ambient temperature, mix of vehicle types, distribution of total vehicles by model year, type of inspection and maintenance program, air conditioning usage, emission control tampering rates, fuel volatility, trip length distribution, and others (USEPA, 1993). Typically, the inputs to Mobile5a are developed for different road classifications (e.g., urban interstate, rural primary arterial) for which the vehicle mix and average speeds may differ.

The emission factor data used to develop the model is based upon laboratory tests for standardized driving cycles (Kini and Frey, 1997). A key limitation of the model is due to the use of driving cycle tests as the basis for model development. For discussion purposes, we will focus upon light duty gasoline vehicles as an example. Data from a total of 11 driving cycles, with average speeds ranging from approximately 2.5 to 65 mph, have been used to develop a "speed correction ratio." Each driving cycle is characterized by an average speed and a specific speed versus time trace. For example, the Federal Test Procedure

(FTP) driving cycle has an average speed of 19.6 mph, but at any instant during the test vehicle speeds range from zero to 57 mph. For some of the high speed driving cycles, speeds vary from zero to 80 mph. Thus, the data used to develop the model represent complete trips, from start to stop, on a particular roadway network. Any given trip may include travel on more than one road class (e.g., feeder/collector street, minor arterial, primary arterial, freeway).

The light duty vehicle emissions are calculated based upon a base emission rate for a new vehicle's emissions on the FTP driving cycle at a 75° F ambient temperature. The emissions are "corrected" for other driving cycles, other ambient temperatures, and other factors (such as for air conditioning usage) by a product of correction factors. Thus, the model assumes that each factor affecting emissions acts multiplicatively.

Although the model was developed using trip-based driving cycles, the model is commonly applied to develop what are termed "link-based" emission inventories. In these inventories, highway networks are divided into highway segments which are termed "links." For each link, an estimate of the daily vehicle miles traveled (VMT) is made. An average speed, typically based upon judgment, is assumed for each link. For example, the average speed on an urban interstate might be assumed to be 57 mph, while the average speed on a rural primary arterial may be 37 mph. These average speeds are entered into Mobile5a and used to develop emission factors. The emission factor based upon the 57 mph average speed is then multiplied by the VMT for urban interstates to estimate the air pollutant emissions for that link.

This approach has several shortcomings, and represents a form of model extrapolation. First of all, the average speed assumed for a particular road class is, in some cases, more nearly a steady cruising speed. In no case is it the average speed for an entire trip, which is the basis of the average speeds associated with the driving cycle emissions data. Therefore, there is a mismatch between the average speed assigned by the user and the average speed underlying the emissions data. Second, the model interpolates between driving cycles by using a "speed correction ratio" based upon average speed. However, driving cycles are not characterized only by average speed, but by the temporal distribution of speeds and accelerations. Two cycles could, in principle, have the same average speed, but very different speed–time profiles and, hence, very different emissions.

There are less obvious limitations to the Mobile5a model. Kini and Frey (1997) studied the data sets and equations which underlie key portions of the model. The purpose of the study was to develop probabilistic estimates of intervehicle variability in emissions and uncertainty in fleet average emissions. Because Mobile5a was developed using point-estimates, no information was readily available regarding the uncertainty in model predictions. In developing a probabilistic version of selected components of the model, Kini and Frey (1997) found that the mean values obtained were typically systematically different than the point estimates of the model. Furthermore, the range of uncertainty in fleet average emission factors was found to be plus or minus 20 to 50%, depending upon the driving cycle and the pollutant considered.

There are additional sources of uncertainty that are difficult to quantify. A key question is regarding the representativeness of laboratory driving cycles with respect to actual on-road vehicle movement. This question has motivated considerable work in alternative methods for characterizing vehicle emissions, including tunnel studies, on-board instrumentation, remote sensing and the development of modal emissions models. The latter are intended to be a more mechanistic approach to predicting emissions based upon engine characteristics and engine loads. All of these approaches require considerable amounts of data, and all are potentially subject to biases due to nonrepresentativeness or failure to capture a full range of vehicle types and operating conditions. These approaches are not directly comparable, making model validation by conventional means difficult, if not impossible. In many cases, it is only possible to obtain qualitative insights regarding model validation. The quest for better mobile source emission factor models is thus an example of many of the issues modelers face regarding verification, interpolation, extrapolation, and validation.

CHAPTER 4

CHARACTERIZING VARIABILITY AND UNCERTAINTY IN MODEL INPUTS

There are many factors affecting the characterization of variability and uncertainty in model inputs that must be considered when developing distributions. Averaging time, spatial coverage, and interindividual variability influence one's selection of a distributional family as well as its parameterization. The following sections outline the assessment of available information about exposure model inputs, the impact of particular sources of variability and uncertainty on the distribution development process (see Chap. 2 for a complete taxonomy), the decision to represent inputs probabilistically, and finally the theoretical basis underlying the selection of some common distributional families.

4.1. DATA AVAILABILITY AND CHARACTERISTICS

The first step in distribution development is to understand the problem being tackled, how the analysis is expected to address the problem, and how its quality will be judged as it is interpreted and used. The nature of the exposure of concern and the type of analysis being done (i.e., back-of-the-envelope, screening or worst-case, typical case, sensitivity analysis, probabilistic analysis) together dictate the data quality objectives. Of course the availability of data may also become a binding constraint.

The second step in distribution development is the assembly of pertinent information. The total body of available information will influence both one's decision to pursue distribution development and one's approach. Information may take the form of empirical evidence about the model input or a surrogate for the input. In the absence of empirical evidence, one's knowledge, reasoning, experience, and subsequent beliefs may be used to construct a subjective distribution. The goals of the analysis and the role of a particular model input will determine in large part the relevance of available empirical data sets. For example, there may be good data reflecting a particular exposure, but these data may not be applicable to other exposure scenarios.

Inputs can be partitioned into four groups on the basis of what is known about them as described in Chap. 1 (see Fig. 1.1). When empirical data are plentiful and relevant to the exposure of concern, distribution development may proceed using statistical techniques. Alternatively, inputs characterized by a complete absence of empirical data are candidates for distributions developed using subjective approaches. Inputs for which there are few or no directly relevant data are candidates for a mixed approach. Two cases occur quite commonly. They include inputs for which there are few directly relevant data and those for which measurements of indirectly relevant or surrogate quantities are abundant. It is rare that empirical information is completely satisfactory, or that none exists.

4.1.1. Temporal Variability

The selection of distributions to represent variability in inputs will depend strongly on the averaging time relevant to the exposure scenario being assessed. Analyses of short-term exposures related to acute health effects must address variability in concentrations, intake rates, and time-activity patterns of exposed populations over many brief intervals. For example, the concentration of PCBs (polychlorinated biphenyls) in ambient air will vary dramatically over minutes and hours depending on many factors, including the status of the source, wind direction, wind speed, and temperature. In contrast, average PCB concentration over days, weeks, or years may be quite stable, as the effects of short-term fluctuations are smoothed at these time scales. The analysis of long-term exposures related to chronic effects would not need to consider the wide range of possible values that would occur in any specific short time period, but rather would focus only on sources of variation that affect long-term average values. Averaging over longer time periods is quite appropriate as long as it is consistent with the assessment question. In Chap. 9 we contrast two approaches: one for developing a distribution for the average PCB concentration in indoor air, where concentration is not subject to significant climatic changes over the year, and the other for developing a distribution relevant to outdoor air, where concentration is affected by weather patterns that change every few days.

Human exposure factors also exhibit a great deal of temporal variability and must be represented with careful attention to their ultimate use. For example, consumption rates of locally grown foods vary with season, and individual physical parameters such as body weight vary with age. Distributions developed to represent such quantities must reflect appropriate time averaging. Long-term averaging, such as on the order of years or decades, may be appropriate when assessing chronic health effects. In contrast, when an acute effect is of concern (e.g., the systemic effect of either carbon monoxide or ozone), model input distributions are developed to represent short term variability, i.e., on the order of hours, days, or months.

Several researchers have explored the complexities of representing long-term variability with data from short-term surveys (Wallace *et al.*, 1994; Price *et al.*, 1996a). This issue requires that long-term distributions of population exposure, which reflect inter-individual heterogeneity, are distinguished from short-term distributions, which reflect seasonal variation, measurement error, and individual activity patterns. Wallace *et al.* (1994) show that measurements of exposure, during multiple time periods, for a fixed population, may be used to partition interindividual variability from temporal variability.

4.1.2. Spatial Variability

Defining the spatial coverage of the exposure scenario under consideration is an important step in the distribution development process. The extent to which spatial averaging is necessary will depend in part on the treatment of other types of variability in model inputs. For example, the size, density, and other distinguishing features of the population of potential receptors, as well as the features of the exposure scenario, will influence the extent to which spatial variability is actually a surrogate for interindividual variability. For example, measurements of the concentration of PCBs in ambient air at ground level receptor locations are variable over large and small geographic regions. This variability arises as a result of the distance between the receptor and the contaminant source, spatial differences in the impact of dilution or transformation processes, the presence of topographic contours and features, and other factors.

Variability is a real feature of the system under study. The question to be answered by an analyst is what level of variability is appropriately associated with the receptor area of concern, rather than how to reduce variability, because only a redefinition of the receptor and/or exposure scenario can bring this about. If available data represent averages on a spatial scale different from the scale of the receptor area, it is necessary to consider the appropriate level of variability to introduce to the analysis. If variability is likely to be under- or overestimated by the analysis the results should include a statement of this possibility and its cause.

4.1.3. Interindividual Variability

It is critical that the population of concern (or multiple subpopulations) be defined precisely for a particular exposure assessment in order to ensure an appropriate treatment of interindividual variability. Characterizing gender, age, ethnicity, occupational status, disease status, and other features of the receptor population leads to insight into appropriate distributions to represent inputs. Survey data on human characteristics must be considered in light of the surveyed population, its relationship to the population of interest, and the duration of the survey. As discussed above, interindividual variability is likely to be greater across short averaging times than across long averaging times (Wallace *et al.*, 1994).

Human physiological characteristics include both those for which empirical evidence is easily collected and thus plentiful (e.g., body weight), providing a good assessment of interindividual variability, as well as those for which interindividual variability is difficult to distinguish from uncertainty due to the limitations of scientific understanding and/or technique. Human behavioral inputs such as, consumption rates for individual food items, are often associated with extremely large interindividual variability.

Aggregation of populations can reduce the amount of variability that must be considered in an analysis. For example, an estimate of the transfer rate of a chemical from a pasture to a cow would depend on whether the analysis was being done for a randomly selected cow at a randomly selected pasture or for an average herd of cows and an average pasture. In the former case, the random combination of a cow and a pasture would be represented by the entire frequency distribution for all cows and all pastures, whereas in the latter case we would only have to consider the sampling distribution of the mean transfer rate, and not the full range of interindividual variability for each cow.

4.1.4. Uncertainty

As detailed in Chap. 2, uncertainty arises from a basic lack of knowledge. Probably the most commonly discussed and easily understood source of uncertainty in exposure model inputs is error associated with measurement and analysis. A second common type of uncertainty stems from a perennial problem plaguing environmental data sets: small sample size. Due to the expense of data collection, there tend to be rather limited numbers of measurements of many exposure model inputs, e.g., environmental concentration, human exposure parameters, and climatological factors. Random sampling error associated with these inputs may become a major contributor to overall variance in exposure or dose (see Chaps. 5 and 9 for examples). A third form of uncertainty arises due to differing views among experts about the interpretation of data or the theoretical bases underlying models.

4.2. DECIDING HOW FAR TO GO

Of all the issues addressed to our expert contributors and peer reviewers in the process of developing this text, one of the most contentious was the question of "whether and how to blend probabilistic analysis with deterministic analysis." For example, should one construct analyses in which some of the model inputs are assigned single point estimates while others are represented by distributions? Sometimes iteration reveals that a simplified exposure model or a subset consisting of the most significant inputs captures much of the overall variability/uncertainty in the exposure estimate (see Chap. 8).

The question to which we apply ourselves throughout this text is actually two-fold in its most reduced form: "when and how?" should probabilistic analyses be carried out. This question encompasses decisions about when probabilistic assessments are useful or necessary (see Chap. 1), but it demands that analysts decide whether, given the full body of available information and knowledge, they are in a position not to proceed with probabilistic analysis. Of course, "how?" is not a solitary question, but rather represents a series of questions, since it is not necessary to assign a distribution to *every* input in an exposure calculation. The question should really be asked in light of the purpose of the assessment, (presumably decision-making). Would this purpose best be served by an assessment in which all model inputs are represented probabilistically or one in which only selected inputs are represented probabilistically?

How does one decide how far to narrow the exposure scenario (characteristics of the receptor, time frame, or spatial coverage) and thus the scope of each random variable requiring description by a distribution? There are many instances in which narrowing the scope of the analysis may be appropriate. For example, in Chap. 9, the estimation of ingestion exposure relies on the development of a distribution describing the mass of locally grown food consumed by adult females who do not work outside the home and who live in a narrowly defined area. The analysis could instead be constructed to separately calculate exposure for two groups of such receptors, one group consisting of those with backyard gardens and the other consisting of those without backyard gardens. Narrowing the receptor definition in this manner could prove important if preliminary analyses indicated the population of greatest concern consisted of those individuals whose rate of consumption of locally grown food was relatively high, and that those individuals supplied themselves through backyard gardens, rather than roadside stands.

Overall, it is our recommendation that the extent and framework of any probabilistic analysis be based on the particulars of the decision supported by the analysis. The foregoing discussion is intended to focus attention on issues regarding sources of variability and uncertainty, in the context of the ultimate purpose of the analysis. We stress that consideration of these issues needs to occur early in the process of thinking about selecting a distributional family. See Section 6.1 for additional caveats on the application of existing distributions in new analyses.

4.3. FAMILIES AND MODELS FOR PROBABILITY DISTRIBUTIONS

A probability distribution model is a description of the probabilities of all possible values in a sample space. A probability model is typically represented mathematically as a probability distribution in the form of either a probability density function (PDF) or cumulative distribution function (CDF).

A random variable is a function defined on a sample space. A sample space is the set of all possible outcomes of the function. A particular outcome is known as a sample point, and is often referred to simply as a sample. A sample space may contain qualitative or categorical information, such as whether a particular measurement is "good" or "bad," or it may contain quantitative information, such as the numerical pollutant concentration at a particular time and location. A sample space that contains a finite number of sample points, or contains what is called a "countable infinity" of samples, is said to be discrete. The latter case refers to a situation in which there may be an infinite number of sample points that have only integer values, such as the number of defects in a piece of material. In contrast, a continuous random variable is one that may take on any value in an interval or perhaps in multiple intervals (Hahn and Shapiro, 1967).

The probabilities associated with a continuous random variable X are determined by the PDF of X. The PDF is a graphical means of representing the relative likelihood with which values of an input may be obtained. The PDF can be denoted $f(x)$ and is defined as follows: $f(x) \geq 0$ for all x. The probabilities associated with discrete random variables are determined by probability distribution functions denoted $P(X = x) \geq 0$ for all x. The probability that x will fall between two numbers a and b is equal to the area under $f(x)$ (or the sum of the probabilities $P(X = x)$) between a and b, and the total area under $f(x)$ (or $P(X = x)$ summed over all x) is equal to 1.

An alternative way to represent a probability distribution is the CDF. The CDF is obtained by integrating (or summing across) the PDF. The y-axis of the CDF is scaled in percentiles (e.g., the 50th percentile or median) and the x-axis shows the value of the distribution associated with each percentile. The CDF shows the fraction of all possible values of x which are less than or equal to a given value of x.

What follows is a discussion of the theoretical basis for selected probability distribution models commonly used in exposure assessment. An understanding of what processes generate certain types of distributions can help with decisions about which distributions to employ to represent specific inputs in exposure models. The special complexity associated with the distributional representation of unknown and variable quantities when data are limited, is addressed in Chap. 5.

4.3.1. Theoretical Basis for Probability Distribution Models

The discussion here is based on more detailed presentations of probability and statistics published by Hahn and Shapiro (1967), Benjamin and Cornell

(1970), Hastings and Peacock (1974), Ang and Tang (1984), Morgan and Henrion (1990), Law and Kelton (1991), Hattis and Burmaster (1994) and others as cited. Illustrations of the probability density functions (and probability distribution functions) of the common distributional forms appear below as Figs. 4.1 through 4.13. Some probability density functions are not explicitly defined for all values of x. These are assumed to be equal to zero outside of the defined range.

4.3.1.1. Lognormal Distribution. One of the most widely used distributional forms in probabilistic assessment is the lognormal. Lognormal distributions have a number of useful characteristics relevant to physical quantities (Aitcheson and Brown, 1957). For example, they assume only non-negative values in the common two parameter form. Also, the lognormal distribution describes random variables resulting from multiplicative processes. Further, the concentration of a chemical in the environment is often well-described by a lognormal because it results from dilution processes in water or air (Ott, 1990; Ott, 1995). Similar reasoning supports the use of the lognormal to represent the degree of deterioration of engineered systems over time. Finally, the lognormal distribution is often used to represent large, asymmetric uncertainties.

The probability density function for the lognormal distribution is:

$$f(x) = \frac{1}{\sqrt{2\pi}\sigma x} \exp\left\{-\frac{[\ln(x) - \ln(m)]^2}{2\sigma^2}\right\}, \quad \text{for } 0 \leq x \leq \infty \quad (4.1)$$

where m is the median or scale parameter ($m = \exp(\mu_{\ln(x)})$) and σ is the standard deviation of $\ln(x)$ or shape parameter (see Fig. 4.1). One common approach to fitting lognormal distributions involves the use of probability plots as discussed in Chap. 5.

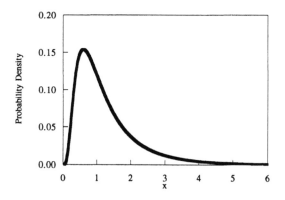

FIGURE 4.1 Lognormal distribution, $m = 1.0$ and $\sigma_{\ln(x)} = 0.6$.

Published examples of exposure model inputs represented by lognormal distributions include contaminant concentrations in environmental media and duration of showering (McKone and Bogen, 1992), human body weights (Brainard and Burmaster, 1992), and the rate at which individuals consume food and soil (McKone and Ryan, 1989; Thompson *et al.*, 1992; Cullen, 1995).

Compared to other probability models, such as the Weibull distribution, the lognormal distribution is "tail-heavy." Although many data sets appear to be lognormally distributed, the lognormal may not be the most appropriate distribution if one is concerned with obtaining a good fit to the upper tail. While the lognormal distribution is an appropriate representation of quantities that arise due to dilution processes as noted above, and thus is widely used to represent concentration data, it is often the case that this is not the only process underlying a particular data set. For example, ambient air pollutant concentrations are based not only upon dilution of air pollutants in the atmosphere, but also upon physical and chemical transformations. Such transformations may affect high concentrations more than low concentrations since many rate-based processes (e.g., chemical reactions) are proportional to the concentration of the pollutants. Thus, processes may occur in addition to mixing that would make the choice of a lognormal distribution inappropriate in some cases. In fact, a less tail-heavy distribution, such as the Weibull, may provide a better fit to the upper percentiles of a particular data set (Seinfeld, 1986).

4.3.1.2. Normal Distribution. The theoretical justification for the normal distribution is the central limit theorem. This theorem states that the distribution of means of independent observations from any distribution, or any combination of distributions, converges to a normal distribution as the number of observations becomes large. Another statement of the theorem is that for samples from a large number of distributions, which may be of any shape, their sum will approach a normal distribution as the number of input distributions increases (as long as no single input contributes substantially to the sum). Thus, when a random variable represents the total effect of a large number of independent "small" causes, the central limit theorem leads one to expect the variable to be normally distributed.

However, the normal distribution suffers from some important limitations. First of all, and this is not often recognized among novice users of probability and statistics, the normal distribution is not a default that can be assumed to apply unless proven otherwise. Many random variables cannot be reasonably regarded as the sum of many small effects, and the burden of proof is upon the analyst to explain why a normality assumption is applicable for a given situation. A second consideration is that the normal distribution has infinite tails. Therefore, when a normal distribution is used to represent a physical quantity which must be non-negative, one must be careful that the coefficient of variation is less than approximately 0.3. If the coefficient of variation is much higher than this, there is a significant probability of predicting negative values. If data analysis indicates a high coefficient of variation, then a normality assumption may be highly inappropriate. This is often true, for example, of trace metal concentrations from power plants, for which there is considerable variability as well as sampling and analytical error (Rubin *et al.*, 1992).

The probability density function for the normal distribution is:

$$f(x) = \frac{1}{\sqrt{2\pi}\sigma} \exp\left\{-\frac{(x-\mu)^2}{2\sigma^2}\right\}, \quad \text{for } -\infty \leq x \leq \infty \qquad (4.2)$$

where μ is the mean and σ is the standard deviation of the random variable X (see Fig. 4.2). In Chap. 5, methods for estimating distributions of these parameters are described.

Empirical evidence suggests that the normal distribution is a good representation for some physical processes and physiological characteristics. Published examples of exposure model inputs which have been represented by the normal distribution include human heights (Brainard and Burmaster, 1992) and the water content of human skin (Thompson *et al.*, 1992).

4.3.1.3. Poisson Distribution. The Poisson distribution is related to the exponential distribution. For events described by the Poisson process (discussed in 4.3.1.4), this distribution describes the number of events occurring within a fixed time interval. The number of events that occur is a discrete value. This distribution can be used, for example, to represent the number of blips on a radiation counter within a specified time interval.

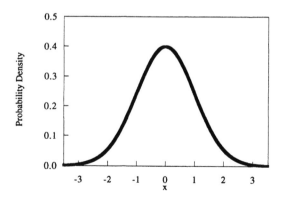

FIGURE 4.2 Normal distribution, $\mu = 0.0$ and $\sigma = 1.0$.

The probability distribution function for the Poisson distribution is:

$$P(X = x) = \frac{\alpha^x \exp(-\alpha)}{x!}, \quad x = 0, 1, 2, 3 \ldots \tag{4.3}$$

where α, the parameter of the distribution, is equal to the average number of events that occur in a time period of fixed length T (see Fig. 4.3). (Note: $\alpha = \lambda T$, where λ is the average rate of occurrence of events and T is the length of the time interval of interest.)

A published example of an uncertain quantity represented by the Poisson distribution is the number of accidents occurring within a fixed time interval (Stinson and Walsh, 1965).

4.3.1.4. Exponential Distribution. The exponential distribution is useful for representing the time interval between successive, random, independent events that occur at a constant rate. Such events are said to occur as a purely random Poisson process. For example, the time between equipment failures, accidents, and storm events can be represented by an exponential distribution.

The probability density function for the exponential distribution is:

$$f(x) = \lambda e^{-\lambda x}, \quad \text{for } 0 \le x \le \infty \tag{4.4}$$

where λ, the parameter of the distribution, is equal to the average rate of occurrence of events, i.e., one divided by the average time between events (see Fig. 4.4).

Lee *et al.* (1995) provide an example of the use of the exponential to represent the concentration of lead in indoor and outdoor air. Also, Shaw and Burmaster (1996) use a variation of the exponential (a Gompertz distribution)

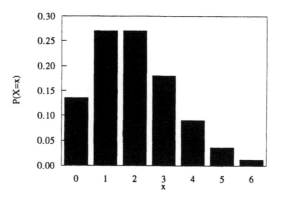

FIGURE 4.3 Poisson distribution, $\alpha = 2.0$.

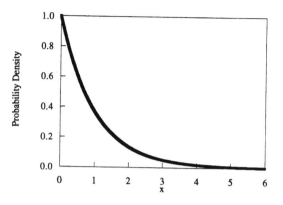

FIGURE 4.4 Exponential distribution, $\lambda = 1.0$.

to represent job tenure for United States workers in selected industries, based on a general analysis of expected residence duration for United States households by Israeli and Nelson (1992).

4.3.1.5. Beta Distribution. The beta is a very flexible distribution, with a finite upper and lower bound (see Fig. 4.5 a-c). Commonly, a two-parameter form of the beta distribution, bounded by 0 and 1, is used to represent judgments about uncertainty. It is useful in Bayesian statistics because the beta can easily be updated to account for new data while retaining prior information (DeGroot, 1986).

The probability density function for the two parameter beta distribution is:

$$f(x) = \frac{x^{\alpha_1-1}(1-x)^{\alpha_2-1}}{B(\alpha_1,\alpha_2)}, \quad \text{for } 0 \leq x \leq 1, \text{ and } \alpha_1, \alpha_2 > 0 \qquad (4.5)$$

where,

$$B(\alpha_1,\alpha_2) = \frac{\Gamma(\alpha_1)\Gamma(\alpha_2)}{\Gamma(\alpha_1+\alpha_2)} \qquad (4.6)$$

and α_1, α_2 are the shape parameters of the distribution (see Fig. 4.5). Note that $\Gamma(x)$ represents the Gamma function, such that,

$$\Gamma(x) = (x-1)! \qquad (4.7)$$

for integer values of x and is tabulated in statistical texts for noninteger values of x (Abramowitz and Stegun, 1965).

The beta has many applications. For example, it can be used to characterize the fraction of time individuals spend engaging in various activities or the

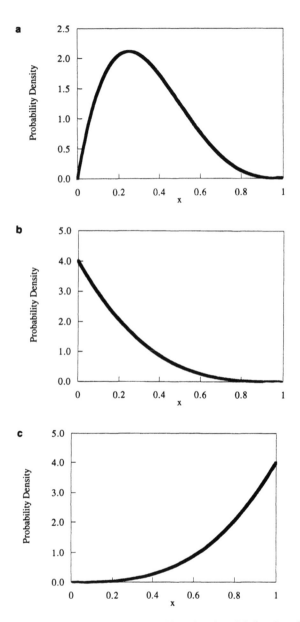

FIGURE 4.5 Beta distribution, a. $\alpha_1 = 2.0$, $\alpha_2 = 4.0$, and scale = 1.0 (i.e., bounded by 0 and 1); b. $\alpha_1 = 1.0$, $\alpha_2 = 4.0$, and scale = 1.0; c. $\alpha_1 = 4.0$, $\alpha_2 = 1.0$, and scale = 1.0.

fraction of time soil is available for dermal contact by humans (as opposed to the time it is covered with snow and ice). Lee *et al.* (1995) used a maximum entropy approach to fit beta distributions to bioavailability factors for metals. The fraction of time an individual spends indoors is represented by a beta distribution in the case study in Chap. 9. A detailed example of the use of the beta distribution to represent partitioning of hazardous air pollutants in a power plant is given in Section 7.6.6.

A 3 or 4-parameter version of the distribution (with minimum and/or maximum values differing from 0 and 1) allows further flexibility for empirical representation. Small and Chicowicz (1986) have published an empirical application of a 3-parameter beta to the reported age of underground storage tanks in Florida. Also, the beta distribution was used to describe concentrations of copper and cadmium in an aquatic system (Bartell and Wittrup, 1996). A general discussion of the beta distribution appears in Benjamin and Cornell (1970) where an application to uncertainty about wind direction appears.

4.3.1.6. Uniform Distribution. The uniform distribution is a special case of the beta distribution which is used frequently in exposure assessment. The uniform is a very simple distribution, requiring an assumption about the range of possible values. For example, the uniform is useful for representing subjective judgments about uncertainty when an expert is only willing/able to estimate an upper and lower bound for a quantity because it is the maximum entropy distribution in this case (see Section 6.2).

The probability density function for the uniform distribution is:

$$f(x) = \frac{1}{b - a}, \quad \text{for } (a \le x \le b) \tag{4.8}$$

where a and b are the minimum and maximum values respectively of the range of the random variable across which the probability density is uniformly distributed (see Fig. 4.6). The mean of a uniform distribution is $(a + b)/2$ and the variance is $(b - a)^2/12$.

A variation of the uniform distribution is the loguniform, useful in cases where inputs cover large ranges of values, but little is known about the shape of their underlying distribution (McKone and Ryan, 1989). In a Loguniform distribution the logtransformed random variable is assumed to be uniformly distributed. The mean of a loguniform is

$$\frac{(\ln(a) + \ln(b))}{2}$$

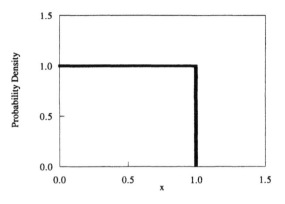

FIGURE 4.6 Uniform distribution, $a = 0.0$, $b = 1.0$.

and the variance is

$$\frac{[\ln(b/a)]^2}{12}.$$

Published examples of the representation of exposure model inputs by the uniform distribution include skin permeability, the ventilation rate of a house, (McKone and Bogen, 1992), soil loading on skin (Thompson, *et al.*, 1992), the ratio of concentrations of indoor and outdoor air (Hawkins, 1991), ingestion rate of soil by cattle (McKone and Ryan, 1989), and the duration of the growing season of plant crops (Cullen, 1995).

4.3.1.7. Triangular Distribution. Rarely, if ever, is there a physical, chemical, or biological process that naturally leads to a random variable which is distributed in the sharp-cornered shape of the triangular. However, the triangular is a maximum entropy distribution used to represent variability and uncertainty when only upper and lower bounds and a most likely value are known (Section 6.2). The triangular distribution also is often used to represent subjective judgments of the maximum value, minimum value, and mode of a random variable.

The probability density function for the triangular distribution is:

$$f(x) = \frac{b - |x - a|}{b^2}, \quad \text{for } a - b \leq x \leq a + b, \tag{4.9}$$

where $a + b$ is the upper bound or maximum and $a - b$ is the lower bound or minimum (see Fig. 4.7).

Triangular distributions may be symmetric or assymmetric. When uncertainties are very large and asymmetric, a logtriangular distribution may be more appropriate (Johnson and Kotz, 1970).

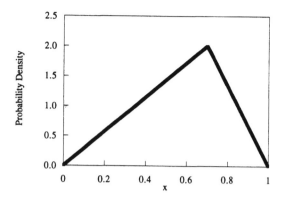

FIGURE 4.7 Triangular distribution, minimum = $a - b$ = 0.0, maximum = $a + b$ = 1.0 and mode = 0.7.

A published example of the representation of an exposure model input by the triangular distribution is the transfer efficiency of a contaminant from water into air (McKone and Bogen, 1992).

4.3.1.8. Weibull Distribution. The Weibull distribution is useful for representing processes such as the time to completion or the time to failure. It is a flexible distribution that can assume negatively skewed, symmetric, or positively skewed shapes. This distribution also may be used to represent non-negative physical quantities. When positively skewed it is often similar in shape to the gamma distribution, but less skewed and with lighter tails. Also, the Weibull is less tail-heavy than the lognormal distribution. When the shape parameter for the Weibull is equal to 1 the distribution is identical to the exponential distribution. A great deal of general information about this distributional form is available (Hahn and Shapiro, 1967; Cox and Oakes, 1984; Crowder *et al.*, 1991; Lee, 1992; and Rao, 1992), as well as applications to environmentally relevant quantities (Seinfeld, 1986). In general, estimation of the parameters of the Weibull distribution from sample data involves the solution of nonlinear equations. A simple graphical procedure for parameter estimation also is available (Hahn and Shapiro, 1967).

The probability density function for the Weibull distribution is:

$$f(x) = \frac{\alpha}{\beta}\left(\frac{x-L}{\beta}\right)^{\alpha-L} \exp\left\{-\left(\frac{x-L}{\beta}\right)^{\alpha}\right\}, \quad \text{for } \alpha, \beta > 0, \, x \geq L \quad (4.10)$$

where β represents the scale parameter, α represents the shape parameter, and L represents the location (see Fig. 4.8).

Published examples of the representation of exposure model inputs by the Weibull distribution include wind speed (Justus *et al.*, 1976), and time between climatic events such as El Nino (Enfield and Cid S., 1991).

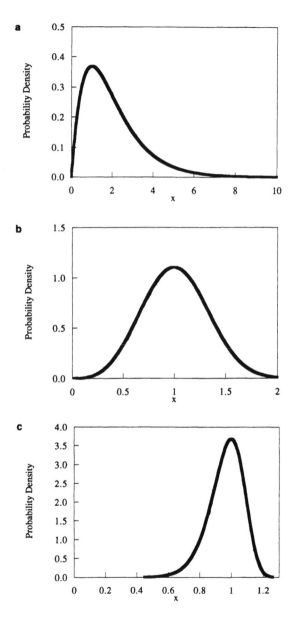

FIGURE 4.8 Weibull distribution, a. $\alpha = 1.0$, $\beta = 1.0$ and $L = 0.0$; b. $\alpha = 3.0$, $\beta = 1.0$, and $L = 0.0$; c. $\alpha = 10.0$, $\beta = 1.0$, and $L = 0.0$.

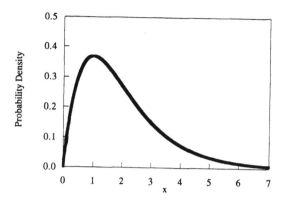

FIGURE 4.9 Gamma distribution, shape = α = 2.0, scale = β = 1.0.

4.3.1.9. Gamma Distribution. The gamma distribution describes the time required for the occurrence of a specified number of events, given a Poisson process. It is useful for representing quantities like the time between events for non-random event processes, the time to complete a task, and the sum of independent exponential random variables. When the shape parameter is equal to 1, the gamma becomes a scalable exponential distribution. A 3-parameter gamma which specifies location is sometimes preferred. General information about the gamma, exponential, and Erlang distributions appear in Cox (1962).

The probability density function for the gamma distribution is:

$$f(x) = \frac{\beta^{-\alpha} x^{\alpha-1} \exp(-x/\beta)}{\Gamma(\alpha)}, \quad \text{for } x, \alpha, \beta > 0 \qquad (4.11)$$

where $1/\beta$ is the scale parameter, α is the shape parameter, and $\Gamma(x)$ is the gamma function, as defined in Section 4.3.1.5 (see Fig. 4.9).

Published examples of the representation of exposure model inputs by the gamma distribution include rainfall properties such as storm amounts, storm durations and time between storms (Eagleson, 1978; Di Toro and Small, 1979; and Small and Morgan, 1986). The gamma has also been used to represent duration of downtime for an offsite power supply for a nuclear power plant (Iman and Hora, 1989).

The Gamma is also useful for representing χ^2, exponential and Erlang distributed quantities in spreadsheet and simulation software packages because these are special cases of the gamma (see Sections 4.3.2.2, 4.3.1.4, and 9.2.2).

4.3.1.10. Gumbel Distribution. The Gumbel distribution, also called "Gumbel's extreme value distribution" or "the extreme value distribution," is

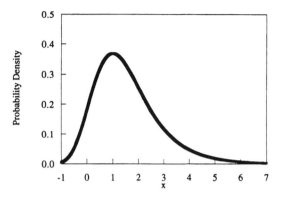

Figure 4.10 Gumbel distribution, L = 1.0 and β =1.0.

frequently used in civil engineering and hydrology. The Gumbel is a distribution of the largest element (maximum value) in a sample, such as, maximum wind gust velocities encountered by aircraft or maximum storm water flow (Hahn and Shapiro, 1967). Additional information about the distribution is available in statistical references such as Hahn and Shapiro (1967); Benjamin and Cornell (1970); Hastings and Peacock (1974); and Kite (1977).

The probability density function for the Gumbel is:

$$f(x) = \frac{1}{\beta} \exp\left[-\frac{x-L}{\beta}\right] \exp\left\{-\exp\left[-\frac{x-L}{\beta}\right]\right\} \qquad (4.12)$$

where β and L are the scale and location parameters respectively (see Fig. 4.10).

The Gumbel distribution has been used to represent maximum annual river flow, a quantity useful in flood analysis, in a range of published work (Kite, 1977; Lettenmaier and Burges, 1982; Resendiz–Carrillo and Lave, 1987).

4.3.1.11. Multimodal or Mixture Distributions. Multimodal, mixture, or custom distributions are useful in a variety of settings (McLachlan and Basford, 1988). Some examples giving rise to multimodal distributions include human characteristics, such as breathing rate and genetic makeup, which vary significantly across different subpopulations (Hattis and Burmaster, 1994). Mixture distributions come in a variety of forms, and are sometimes used to address issues of surprise (Section 6.4). One type of mixture encompasses multiple distributional families in a single probability density function, such that one represents the tail and another the main body of the distribution. Candidates for the use of this type of mixture distribution include weather phenomena such as extremes of rainfall

TABLE 4.1 Example of Chance Distribution

x	P(x)
0.20	0.10
0.40	0.20
0.60	0.30
0.90	0.30
1.00	0.10

or wind (Lambert *et al.*, 1994). A second type of mixture distribution consists of a probability density function in contiguous pieces, each from the same distributional family but with different parameters. A third type of mixture involves different distributional families (discrete, continuous, or both). For example, Frey and Rhodes (1996) represent plant capacity utilization by a mixture distribution involving a beta, a normal, and a delta function assuming the value zero. This example and the implementation of mixture distributions in a Monte Carlo simulation is discussed in Section 7.5.1.3.

4.3.1.12. Fractile and Chance Distributions. Fractile and chance distributions are often used in expert elicitation or to represent empirical data without any additional judgments regarding distributional shape. In the fractile distribution, the finite range of possible values is divided into subintervals. Within each subinterval, values are sampled uniformly according to a specified frequency (see Fig. 4.11). This distribution can be used to represent any data or judgment about uncertainties in a continuous input. The chance distribution resembles the fractile distribution, except that it applies to discrete inputs (see Table 4.1 and Fig. 4.11).

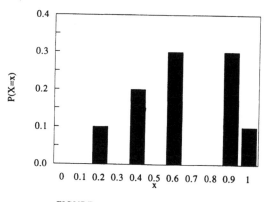

FIGURE 4.11 Chance distribution.

4.3.2. Characterizing Uncertainty in Distribution Parameters

Often a particular distributional form is selected to represent variability in an input on the basis of some underlying theory or mechanism known to the analyst. In such cases the challenging task of estimating the parameters of that distribution remains. Two sampling distributions are particularly useful for representing uncertainty regarding the parameters of a frequency distribution for variability: the Student's t distribution, and the chi-square (χ^2) distribution.

Specifically, when based on randomly sampled data sets of limited size, sample statistics such as the sample mean, $\bar{x}$, and sample variance, s_x^2, are estimates of the true but unknown values of the population parameters. These estimators themselves are described by *sampling distributions*. In order to create such a sampling distribution, a *parent distribution* of a random variable X is sampled n times $\bar{x}$, and s_x^2, are calculated (see Chap. 5 for more details). Multiple rounds of sampling from the parent distribution yield a distribution of sample means, variances, or other statistics. These sampling distributions for a data set of sample size n often have a characteristic shape. Examples of this include the Student's t distribution for describing the sample mean and the χ^2 distribution for the sample variance of a normal distribution. As the focus of probabilistic approaches to exposure assessment shifts toward two dimensional representations of exposure (i.e., uncertainty in one dimension and variability in the other) there is likely to be increased use of these two distributions (see Chaps. 5, 7 and 9, as well as, Frey and Rhodes, 1996).

4.3.2.1. Student's t Distribution. The Student's t distribution is the sampling distribution for the population mean of a normal random variable estimated from a finite number of samples. Theoretically, a derived distribution of sample means from a normally distributed parent population is described by a Student's t dis-

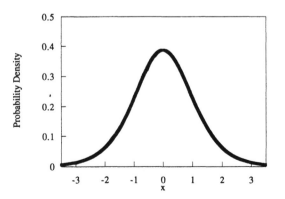

FIGURE 4.12 Student's t distribution $\upsilon = 9$, illustration is restricted to values from −3.0 to +3.0.

tribution. With a large number of samples (sample size = n), the sampling distribution asymtotically becomes normal with a mean equal to the mean of the parent distribution, and a variance equal to $1/nth$ of the variance of the parent distribution. In fact, as the sample size increases, even a non-Normal parent distribution yields a normally distributed population mean distribution.

The probability density function for the Student's t distribution is:

$$f(x) = \frac{\Gamma\left[\frac{(\upsilon + 1)}{2}\right]\left[1 + \frac{x^2}{\upsilon}\right]^{-\frac{(\upsilon+1)}{2}}}{\sqrt{(\pi\upsilon)}\Gamma(\upsilon/2)}, \quad \text{for } -\infty \le x \le \infty \quad (4.13)$$

where υ represents the number of degrees of freedom, which is equal to $n - 1$ (see Fig. 4.12).

See Chap. 9 for an example involving a mean represented by a Student's t distribution using commercially available software.

4.3.2.2. Chi-square Distribution (χ^2). The chi-square distribution represents the sum of several independent squared standard normal random variables! This distribution can be used to represent uncertainty in the sample variance from a normally distributed parent population. Sample variance is distributed χ^2 with $\upsilon = n - 1$ degrees of freedom, where n is the number of samples from the parent distribution used in the calculation. The χ^2 distribution also is used to describe the sum of the squared error of the differences between model predictions and observations of a quantity as described in Section 5.8.1. The χ^2 distribution is a special case of the gamma distribution, with $\beta = 2$ and $\alpha = \upsilon/2$.

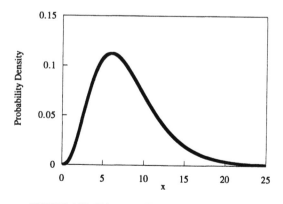

FIGURE 4.13 Chi-square distribution, $n = 9$, $\upsilon = 8$.

Frequencies associated with values of the χ^2 are available in tabular form in statistics texts (Hahn and Shapiro, 1967) and some values appear in Table 5.3. The χ^2 distribution with υ degrees of freedom takes the form:

$$f(x) = \frac{x^{\left((\upsilon-2)/2\right)} \exp(-x/2)}{2^{\upsilon/2}\,\Gamma(\upsilon/2)} \tag{4.14}$$

where $\Gamma(\upsilon/2)$ is the gamma function with parameter $\upsilon/2$ (see Fig. 4.13).

See Chap. 9 for an example of the representation of a variance with a χ^2 distribution using commercially available spreadsheet-based simulation software.

CHAPTER 5

DATA AND DISTRIBUTIONS

In this chapter, we consider a number of issues related to the development, interpretation, and refinement of probability distributions that are used to quantify variability and uncertainty in exposure model inputs.

We present some statistical techniques for characterizing distributions based upon analysis of data sets. First, we describe in detail a number of summary statistics that are useful in probabilistic analysis, some of which were introduced in Chap. 4. Summary statistics include quantities such as the mean, variance, and percentiles of a distribution. Typically, these statistics are estimated using relatively small sets of sample data. Thus, there is uncertainty in the estimates of these statistics due to sampling error. When the data are a representative random sample, the quantification of sampling error may be done using classical statistical techniques or numerical simulation methods. These techniques are discussed. If the data are nonrepresentative or not a random sample, then additional uncertainty may exist. In such cases, it may be necessary to use expert judgment, in addition to data analysis, to more fully quantify the uncertainties. Methods for quantifying uncertainty using expert judgment, obtained via formal elicitation protocols, are briefly mentioned in Chap. 6 and are discussed elsewhere in more detail (e.g., Kahneman *et al.*, 1982; Morgan and Henrion, 1990; NCRP, 1996).

Some general features of probability distribution models, also referred to as probability models, are discussed in Section 5.6. Methods for making inferences regarding the selection of probability models based upon analysis of data are discussed in Sections 5.6 through 5.8. These sections deal with the use of the method of moments, maximum likelihood, probability plots, and goodness-of-fit tests to select probability models and make estimates of their parameters. In addition, the use of empirical distributions is discussed. This book is primarily concerned with continuous random variables, such as would be used to describe intake rates, concentrations, exposure durations, averaging times, body weights, and other continuous quantities used in exposure assessment.

5.1. SUMMARY STATISTICS

Often, one wishes to use only a few descriptive values to summarize infor-mation about the distribution of a random variable. Three of the key character-istics of a distribution are its central tendency, dispersion, and shape. There are several measures of central tendency. These include the mean, median, and mode. The dispersion, or spread, is measured by the variance of the distribution. The shape of a distribution is also reflected by measurable quantities, which include skewness and kurtosis. However, in the latter case, we may be more interested in the qualitative description of the distribution (e.g., skewed, multimodal) than the value of quantities which may reflect these characteristics.

5.1.1. Central Tendency

There are several measures of central tendency. In this section, we expand upon the discussion in Chap. 4 in defining the mean, median, and mode of a dis-tribution. These three measures of central tendency are compared. Then we il-lustrate, using the example of the leafy produce PCB concentration data set from Chap. 9, how these measures of central tendency may be inferred from an em-pirical data set.

5.1.1.1. Definitions of the Mean, Median, and Mode. There are several names for one of the most useful measures of central tendency: expected value, arithmetic mean, average, or mean. The mean can be regarded as the center of gravity of a probability density function, and therefore is sometimes referred to as the "centroid." Thus, the mean is the value of a random variable for which the weighted probability mass for all values less than the mean is equal to the weighted probability mass for all values greater than the mean. The weighting is based upon the distance of a value of the random variable from the mean. For a discrete distribution, the mean is calculated as follows:

$$E(x) = \sum_{i=1}^{n} x_i p_i \qquad (5.1)$$

where $E(x)$ is the expected value, x_i is a sample point for the random variable X, and p_i is the probability of obtaining the ith sample point. For a continuous ran-dom variable, the summation is replaced by an integral:

$$E(x) = \int x \, f(x) \, dx \qquad (5.2)$$

where the integral extends over the entire range of values of x. In this case, x is any value of the random variable X and $f(x)$ is the probability density function of x. The mean is referred to as the first moment of a distribution. Eq. (5.2) produces the first moment of the distribution with respect to the origin. Higher order moments defined with respect to the mean are referred to as "central" moments. In general, the kth central moment for any continuous distribution is:

$$\mu_k = E[x - \mu_1]^k \int (x - \mu_1)^k f(x) \, dx \qquad (5.3)$$

where μ_k is the kth moment about the mean, μ_1. The second central moment is the variance. The third and fourth central moments are related to skewness and kurtosis, respectively, which are measures of the shape of the distribution. Higher order central moments are rarely used in practice. A distribution is completely specified once all of its central moments are known. For example, a normal distribution is completely specified by the mean and the second central moment.

Unlike other measures of central tendency, the mean has special properties. For example, the central limit theorem implies that the mean of a sum of independent random variables for which no individual variable dominates the sum is approximately a normal distribution with a mean equal to the sum of the means and a variance equal to the sum of the variance. This is discussed further in Section 7.1.

Another measure of central tendency is the median, also known as the midpoint or 50th percentile of the distribution. The median is the point such that exactly half of the probability is associated with values less than the median and half of the probability is associated with values greater than the median.

A third measure of central tendency is the mode. The mode is sometimes referred to as the "most likely" value. The mode of a continuous random variable is the value associated with a maximum of the probability density function. Some distributions may have more than one mode. In such cases, there is a mode at each local maximum of the probability density function.

5.1.1.2. Comparisons of the Mean, Median, and Mode. The mean, median, and mode are the same for unimodal, symmetric distributions, such as the normal distribution shown in Fig. 5.1(a). These three measures of central tendency may differ significantly for asymmetric distributions, as illustrated in Fig. 5.1(b). Generally, for unimodal, positively skewed distributions, such as the lognormal, the mean will be larger than the median, and the median will be larger than the mode. In extreme cases, it is possible for the mean to be above the 90th percentile of the distribution. The median is sometimes a more robust measure of cen-

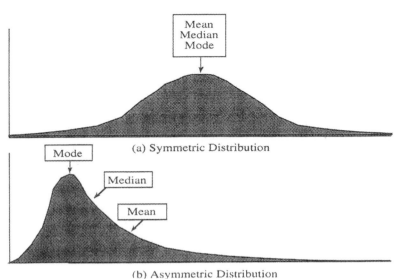

(b) Asymmetric Distribution

FIGURE 5.1 Relative location of mean, median, and mode for symmetric and positively skewed unimodel distributions.

tral tendency, because it is not sensitive to the shape of the tail of a distribution, or to the occurrence of outliers in a data set, as is the mean.

5.1.1.3. Example Based upon Leafy Produce PCB Concentration Data Set.

The mean, median, and mode can be estimated from observed data. Suppose that we have some number, n, of observations of sample values (or data points) which we believe are drawn from some underlying probability distribution for a random variable, X. Then we can denote each observation as x_i, where i varies from 1 to n. The mean of the data is given by:

$$\bar{x} = \frac{\Sigma_{i=1}^{n} x_i}{n} \tag{5.4}$$

For example, the mean of the leafy produce PCB concentrations given in Table 9.3 of Chap. 9 is 0.22 ng PCB/g of produce, based upon an average of nine samples.

The quantity $\bar{x}$ refers to an estimate of the expected value $E(X)$. The mean of the data set is only an estimate of the mean of the underlying probability distribution from which the data were observed. If we had more data points, or had sampled a different set of data from the same distribution, we would have obtained a different estimate for the mean. However, as we obtain an increasingly large data set, the estimate of the mean should converge to the true value of the mean. Thus, we can estimate a confidence interval for the mean, which is a function of the number of data points available. We will return to this topic in Section 5.5.

The median of the data set is determined by rank ordering all of the observed data according to the magnitude of each data point. If the n observations are ordered from smallest ($i = 1$) to largest ($i = n$), then the median, $x_{0.50}$, is given by:

$$
x_{0.50} = \begin{cases} \text{value of the } \left(\dfrac{n+1}{2}\right)^{\text{th}} \text{ observation, if } n \text{ is odd} \\[2em] \text{midpoint between } \left(\dfrac{n}{2}\right)^{\text{th}} \text{ and } \left(\dfrac{n}{2}+1\right)^{\text{th}} \text{ observation, if } n \text{ is even} \end{cases} \tag{5.5}
$$

For example, the median of the leafy produce PCB concentration in Table 9.3 of Chap. 9 is 0.25 ng PCB/g of produce, which in this case is higher than the mean.

The mode of a data set is sometimes difficult to estimate reliably. Often, one groups the data into bins of some selected interval and counts the number of data points contained in each bin. In this situation, the mode is usually defined as the center point of the bin that contains the most data points. In the case of the leafy produce PCB concentration example, the mode appears to be 0.28 ng PCB/g of produce. The estimate of the mode is sensitive to the size and placement of the bins, and is also affected by random sampling error.

5.1.2. Spread

The variance of a distribution is a measure of its spread or dispersion. The variance of a continuous distribution is given by the second central moment with respect to the mean:

$$
\sigma^2 = \mu_2 = E[x - \mu_1]^2 = \int (x - \mu_1)^2 f(x) \, dx \tag{5.6}
$$

An estimate of variance is referred to as s^2, rather than as σ^2. The latter refers to the true value of the variance which would be known only if an infinite number of samples were available. The variance may be estimated from a data set using the following equation:

$$s^2 = \frac{1}{n} \sum_{i=1}^{n} (x_i - \bar{x})^2 \tag{5.7}$$

For small sample sizes, Eq. (5.7) is said to be biased. The value of the mean used in the equation for estimating the variance is itself only an estimate. It is not the true value of the mean, which can only be known if an infinite amount of data are collected. Since the data set is used to estimate the mean, there are only $(n-1)$ degrees of freedom remaining for estimating the variance. Therefore, the unbiased estimate of variance is:

$$s^2 = \frac{1}{n-1} \sum_{i=1}^{n} (x_i - \bar{x})^2 \tag{5.8}$$

Of course, the difference between the biased and unbiased estimators becomes negligible as n becomes large.

The square root of the variance is known as the standard deviation. A non-dimensional measure of the spread of a distribution is the standard deviation divided by the mean, which is known as the coefficient of variation, v:

$$v = \frac{\sigma}{\mu} \tag{5.9}$$

The standard deviation has useful significance in making estimates of probability bands for many standard probability distributions. For example, for a normal distribution, a range of ± 1.65 standard deviations from the mean encloses approximately 90% of the values of the distribution, while a range of ± 1.96 standard deviations encloses approximately 95% of the values. Probability bands calculated based on the assumption of a normal distribution are invalid when data cannot be described by a normal distribution. Thus, for example, you cannot use these ranges to calculate probability bands for lognormal, gamma, beta, Weibull, or other types of distributions. Instead, you can use the appropriate CDF to determine the end points of any desired probability range.

Other measures of spread include probability ranges and the interquartile range. A probability range is any arbitrary range encompassing a specified fraction of samples from a data set. For example, a 90% probability range is typically bounded by the 5th and 95th percentiles of a CDF. An interquartile range is a more specific variant of a probability range. It is the range of values enclosed by the 25th and 75th percentiles.

5.1.3. Shape

There are several characteristics of the shape of a continuous probability distribution. These include skewness, kurtosis, and the number of modes. In addition, we consider how information regarding the skewness and kurtosis can be used to select a parametric probability distribution to represent the data.

Skewness is the asymmetry of a distribution. For example, in Fig. 5.1, the normal distribution is symmetric and has a skewness of zero. The lognormal distribution is positively skewed, with a long tail toward the positive side of the abscissa. Skewness is based upon the third central moment of the distribution. For a continuous distribution, the third central moment is given by:

$$\mu_3 = E[x - \mu_1]^3 = \int (x - \mu_1)^3 f(x) \, dx \qquad (5.10)$$

The third central moment may be estimated from a data set using the following relation:

$$m_3 = \frac{\sum_{i=1}^{n}(x_i - \bar{x})^3}{n} \qquad (5.11)$$

Bias corrections are available for the estimator of the third central moment (Bobee and Robitaille, 1975).

The skewness, γ_1, is commonly defined relative to the degree of spread of the distribution, and is given by the third central moment divided by the cube of the standard deviation:

$$\gamma_1 = \frac{\mu_3}{\sigma^3} \qquad (5.12)$$

Thus, skewness may have values that are positive, negative, or zero. For quantities that must be nonnegative, such as concentrations, intake rates, exposure durations, and many other exposure parameters, it is common to have positively skewed distributions that reflect variability. Random measurement errors, on the other hand, may commonly have no skewness. As examples, a normal distribution has a skewness of zero, while a lognormal distribution has a positive skewness that is directly proportional to the coefficient of variation.

For some purposes, the square of the skewness is used. This quantity is known as β_1 and is given by:

$$\beta_1 = \left(\frac{\mu_3}{\sigma^3}\right)^2 \tag{5.13}$$

Thus, the skewness is sometimes represented as $\sqrt{\beta_1}$.

Kurtosis refers to the peakedness of a distribution. Kurtosis is estimated based upon the fourth central moment of the distribution. The fourth moment of a continuous distribution is given by:

$$\mu_4 = E[x - \mu_1]^4 = \int (x - \mu_1)^4 f(x) \, dx \tag{5.14}$$

The fourth central moment may be estimated from a data set using the following relation:

$$m_4 = \frac{\sum_{i=1}^{n}(x_i - \bar{x})^4}{n} \tag{5.15}$$

The kurtosis is a nondimensional quantity. It is the fourth moment of the distribution relative to the square of the variance:

$$\gamma_2 = \frac{\mu_4}{(\mu_2)^2} = \frac{\mu_4}{\sigma^4} \tag{5.16}$$

The kurtosis is also referred to as β_2. A flat distribution, such as the uniform distribution, has a lower kurtosis than a highly peaked distribution, such as the normal or lognormal distributions. Kurtosis is perhaps less commonly used than skewness. However, both measures can be useful when trying to select probability distribution models to fit to data sets, as discussed in the next section. Many standard computer packages, including spreadsheets such as the Microsoft Excel program, have built-in functions to calculate these quantities.

The number of modes of a distribution is also of interest. Typically, though perhaps not always, distributions that reflect uncertainty have only a single mode. However, distributions that reflect variability may have several modes, which indicate the presence of subgroups of the population that share some characteristics unlike those of other members of the population. The observation of multimodal data is an indication that there may be subgroups that should be treated separately in an exposure assessment. A multimodal distribution can often be decomposed into a mixture of component distributions, each of which has only one mode. A variety of techniques, such as maximum likelihood estimation, probability plotting, and goodness-of-fit tests, can be used to ascertain the component distributions of a mixture (McLachlan and Basford, 1988; D'Agostino and Stephens, 1986).

5.2. EMPIRICAL BASIS FOR SELECTING A PARAMETRIC PROBABILITY DISTRIBUTION MODEL

The skewness and kurtosis are useful in helping to select an appropriate parametric distribution to fit a data set. Fig. 5.2 displays the relationship between the square of the skewness (β_1) and kurtosis (β_2) of many standard parametric probability distributions. The normal distribution always has a skewness of zero and a kurtosis of three and, therefore, is represented as a point on the skewness–kurtosis plane. In the case for which a lognormal distribution has a very small coefficient of variation, the skewness is close to zero, and it will appear similar to a normal distribution. As the coefficient of variation of the lognormal increases, so will its skewness, and thus will deviate from normality.

Like the normal distribution, the uniform and exponential distributions are also represented by single points on the β_1-β_2 plane. These three distributions have only one shape. The other distributions illustrated in the figure can take on different shapes and, therefore, appear as lines or regions on the plane.

The lognormal and gamma distributions have very similar curves on the β_1-β_2 plane. These two distributions are very similar for moderate or low skewness and, therefore, can often provide equally good (or bad) fits to data. As the skew-

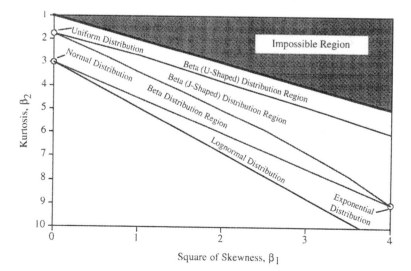

FIGURE 5.2 β_1-β_2 Plane depicting several common parametric probability distributions.

ness increases, the kurtosis of these two distributions will diverge further. Although not shown in the figure, the Weibull distribution can also take on shapes similar to the lognormal and gamma. Thus, it is often the case that lognormal, gamma, or Weibull distributions may adequately fit a data set. The beta distribution can take on a variety of shapes, including both "U" and "J" shapes for the probability density function, and therefore is represented by areas, rather than lines, in the plane. The impossible region represents combinations of skewness and kurtosis that are theoretically impossible.

Other systems of distributions have been developed, the purpose of which have been to more fully realize possible combinations of skewness and kurtosis. For example, the Johnson system of distributions is depicted in Fig. 5.3. The Johnson distributions include: (a) the Johnson S_L distribution, which is a three-parameter lognormal distribution that includes a location parameter in addition to the scale (mean) and shape (standard deviation) parameters; (b) the Johnson S_B distribution family, which is a four-parameter distribution for random variables bounded from both above and below; and (c) the Johnson S_U distribution family, which is a four parameter distribution for unbounded random variables. The four-parameters of the latter two families include two shape parameters, one location parameter, and one scale parameter. Many textbooks, such as Hahn and Shapiro (1967), provide a full treatment of the Johnson system of distributions. This system of distributions offers more flexibility for representing an empirical data set than do the standard parametric distributions given in Fig. 5.2.

A system of distributions was also proposed by Pearson. The Beta distribution is a Pearson Type I distribution. The gamma distribution is a Pearson Type

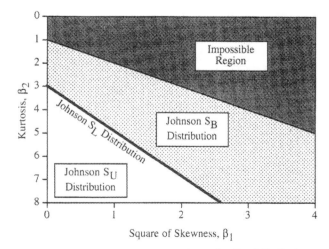

FIGURE 5.3 β_1-β_2 Plan depicting the Johnson Family of distributions.

III distribution. For more information on the Pearson system of distributions, the reader is referred to Hahn and Shapiro (1967).

As an example, the mean, selected other central moments, and selected other statistics of three data sets given in the case study of Chap. 9 are presented in Table 5.1. The data are for the concentration of PCBs in leafy, vine, and root vegetables (see Table 9.3). The data are purposefully reported with no more than two digits. In fact, in some cases this may imply far more precision than actually exists regarding estimates for small data sets. We will return to this in more detail when we consider confidence intervals for the statistics of data sets.

We have used both the "biased" and "unbiased" estimators of standard deviation. For the smallest data set, we observe the largest difference between these two. As the number of data points increases, the difference between the two estimators becomes smaller.

The coefficient of variation provides an indication of the relative dispersion of the data. In all cases, the coefficient of variation is greater than 0.4. This implies that it would be inappropriate to use a normal distribution to describe these data. For example, the mean of the root vegetable PCB concentration data is only 1.53 standard deviations greater than zero. If a normal distribution were used to represent these data, there would be a 6.3% probability of predicting physically impossible negative values for this quantity. As a rule of thumb, when the mean is within approximately three standard deviations of zero, it is typically inappropriate to use a normal distribution for a quantity that cannot have negative values. Of course, there may be cases in which a truncated normal distribution would be appropriate.

As noted previously, the leafy vegetable data are negatively skewed, and this is confirmed by the negative value of the coefficient of skewness. As ex-

TABLE 5.1 Examples of Summary Statistics for Three Data Sets of Appendix A: Concentration of PCB in Produce

	Data set		
Description	Leafy Produce	Vine Produce	Root Produce
Number of data points	9	17	25
Mean, ng/g	0.22	0.10	0.19
Second central moment (variance), $(ng/g)^2$	0.0088	0.0022	0.016
Standard deviation, ng/g	0.094	0.047	0.13
Coefficient of variation	0.42	0.47	0.65
"Unbiased" standard deviation, ng/g	0.099	0.048	0.13
Third central moment, $(ng/g)^3$	-4.7×10^{-4}	1.8×10^{-4}	2.5×10^{-3}
Skewness	-0.58	1.7	1.2
Fourth central moment, $(ng/g)^4$	1.7×10^{-4}	3.2×10^{-5}	9.4×10^{-4}
Kurtosis	2.2	6.5	3.7

pected, the vine and root data sets are positively skewed. The kurtosis of the three data sets varies over a range of approximately two to seven. These values of kurtosis indicate that the distributions may have strong peaks.

Based upon the skewness and kurtosis, we might consider selecting a parametric distribution to represent these data. It would appear, for example, that for the leafy vegetable data, if we assume that the observed skewness is not significantly different than zero, then perhaps a symmetric distribution such as a normal or a uniform would provide a compatible value for the estimated kurtosis. For that matter, a lognormal distribution may also be a reasonable approximation, although the selection of the lognormal distribution imposes a non-negative skewness upon our representation of the data. For the vine produce data, the skewness and kurtosis appear to be a nearly exact match for a lognormal distribution. For the root produce data, the skewness and kurtosis appear to be within the region for the Beta distribution. Thus, one might hypothesize, in the absence of other information or knowledge, that the following distributions should be used for each data set: normal or uniform for the leafy vegetables; lognormal for the vine produce; and beta for the root produce. However, we have good reason to suspect that a normal distribution is inappropriate for the leafy vegetable data, as previously discussed. Clearly, some additional exploration and judgment is needed.

Another important factor to consider is that with such small sample sizes as for our example data sets, there is a substantial amount of uncertainty in the estimates of the skewness and kurtosis. Thus, one should not place absolute faith in the validity of the point estimates for these statistics.

5.3. EMPIRICAL CUMULATIVE DISTRIBUTION FUNCTIONS

Data are often summarized using empirical cumulative distribution functions. These functions provide a relationship between fractiles and quantiles. A fractile is the fraction of values that are less than or equal to a given value of a random variable. A quantile is the value of a random variable associated with a given fractile. The term percentile is often used in place of fractile. The median is an example of a fractile. The median has a fractile of 0.5 or a percentile of 50%. The value of the random variable at the median is the quantile at the 50th percentile. Often, we are interested in knowing the range enclosed by the 0.01 and 0.99 fractiles, which provide an indication of the dispersion of the distribution. In exposure assessments, we are also interested in knowing what portion of a population faces an exposure less than or equal to some level, such as the 90th or 95th percentile.

5.3.1. Plotting Position

To estimate a fractile or percentile from data requires rank ordering the data. We will consider several possible methods for estimating the percentile of an empirically observed data point. These methods are referred to as "plotting positions." The plotting position is an estimate of the cumulative probability of a data point. Harter (1984) provides an overview of the various types of plotting positions.

A general expression for many of the possible plotting positions was suggested by Blom (1958):

$$F_X(x_i) = Pr(X < x_i) = \frac{i - \alpha}{n - 2\alpha + 1},$$

$$\text{for } i = 1, 2, \ldots, n \text{ and } x_1 < x_2 < \cdots < x_n \qquad (5.17)$$

where $Pr(X < x_i)$ denotes the probability that the random value X will have values less than that of the sample x_i. For Eq. (5.17), the samples of X are ranked in ascending order, and the index i refers to the rank of each sample.

Inherent in the development of estimators for plotting positions is that there is a sampling distribution for any given percentile due to random sampling error. The mean plotting position is given by the following (Gumbel, 1958):

$$F_X(x_i) = Pr(X < x_i) = \frac{i}{n + 1},$$

$$\text{for } i = 1, 2, \ldots, n \text{ and } x_1 < x_2 < \cdots < x_n \qquad (5.18)$$

This is a special case of Blom's expression in which $\alpha = 0$. A benefit of using the quantity $(n + 1)$ in the denominator is that the smallest and largest values in the samples are not given percentiles of 0 and 1, respectively.

The modal plotting position is given by:

$$F_X(x_i) = Pr(X < x_i) = \frac{i - 1}{n - 1},$$

$$\text{for } i = 1, 2, \ldots, n \text{ and } x_1 < x_2 < \cdots < x_n \qquad (5.19)$$

Hazen (1914) suggested the following plotting position:

$$F_X(x_i) = Pr(X < x_i) = \frac{i - 0.5}{n},$$

$$\text{for } i = 1, 2, \ldots, n \text{ and } x_1 < x_2 < \cdots < x_n \qquad (5.20)$$

Using Blom's suggested framework, Hazen's plotting position, with $\alpha = 0.5$, appears to be a compromise between the mean plotting position ($\alpha = 0$) and the modal plotting position ($\alpha = 1$) (Harter, 1984).

With all of these variants on the plotting position, and many others which we have not listed, which one should be used? Gumbel (1958) suggested some postulates that provide insight into the selection of plotting position. These include: all observations should be able to be plotted; the observations should be equally spaced on the frequency scale; the plotting position should be independent of the distribution chosen to represent the data; and the plotting position ought to be analytically simple. The latter is clearly a matter of judgment. In addition, Gumbel suggested there should be criteria regarding the likelihood of obtaining sample values larger or smaller than the minimum and maximum values in a data set of size n. Gumbel's five criteria are not universally accepted (Harter, 1984). For example, some do not agree with the notion that data should be equally spaced on the frequency scale. Also, it may not be possible to have a truly distribution-free plotting position that will yield good results for all possible applications. In fact, the choice of an appropriate plotting position can depend on the underlying distribution of the data and the purpose of the analysis. Thus, there may be reasons to prefer different plotting positions for testing the fit of a distribution or for extrapolating extreme values of a distribution. Some plotting positions have known deficiencies. For example, for small sample sizes the mean plotting position leads to overestimates of the standard deviation, while the Hazen plotting position leads to underestimation of the standard deviation. As the sample size increases, the various plotting positions tend to approach each other and, therefore, differences between them become less important.

For the most part, many investigators use either the mean or Hazen plotting positions. In some situations, such as with the use of transformations for constructing probability paper, a specific plotting position may have been assumed and thus must be used in order for the probability paper to be valid. Thus, you should check whether a particular plotting position is specified when using transformations or goodness-of-fit tests.

5.3.2. Examples of Empirical Cumulative Distributions

In Fig. 5.4, empirical CDFs are plotted for six hypothetical data sets. The Hazen plotting position was used in these examples. Each of these six data sets

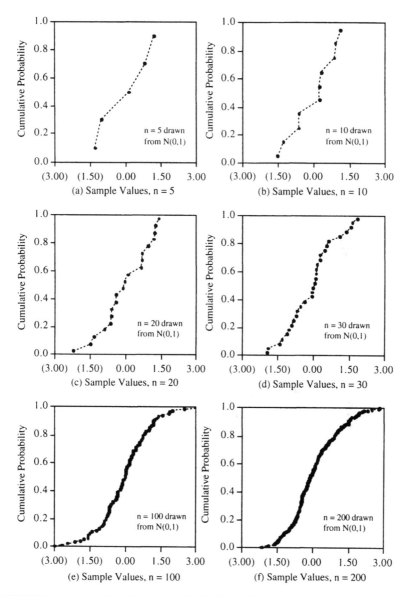

FIGURE 5.4 Examples of empirical cumulative distribution functions for six hypothetical data sets.

involves samples drawn at random from a standard normal distribution, which has a mean of zero and a standard deviation of one. The six data sets differ in the number of samples that were drawn, ranging from five to 200. These six graphs illustrate that with small sample sizes, it may be difficult to discern the underlying distribution, while even with a relatively large number of random samples (e.g., 200) the representation of the tails may be poor. For example, in Fig. 5.4(a), there are only five data points. In this case, it would be difficult to discriminate among a large variety of possible distributions that might represent these data (e.g., normal, lognormal, gamma, Weibull, uniform). In Fig. 5.4(f), there are 200 data points. However, even here there is some ambiguity regarding the appropriate distributional model. This is because it appears that the lower tail of the distribution is not as well represented as is the upper tail. Specifically, the distribution does not appear to be symmetric. If it were perfectly symmetric, there would be as many values below –2 as there are values above +2. These results emphasize that even for data that meet stringent assumptions regarding independence and random sampling, there can be random fluctuation in the sample data that may make it difficult to select the "true" distributional model that best represents the data.

The effect of scatter in the data becomes more apparent when we look at some actual data. In Figs. 5.5 through 5.7, data regarding the concentration of PCBs in three types of vegetables are plotted. These data are taken from Table 9.3 of Chap. 9. In the first case, there are only nine data points, and it is difficult to identify what type of probability distribution model might be appropriate to represent these data. In the other two cases, it seems apparent that the data are positively skewed and, therefore, might be described by distributions such as the lognormal, gamma, or Weibull. We will return to these three examples as we consider other statistical techniques for analyzing data.

5.4. OTHER WAYS TO VISUALIZE DATA

While the empirical CDF is a convenient way to visualize data, there are other ways that some analysts or decision makers may find to be more useful.

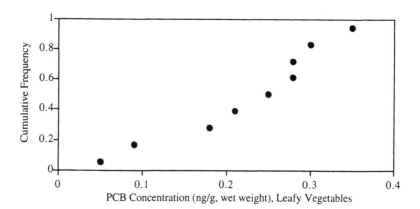

FIGURE 5.5 Empirical distribution of leafy produce PCB concentrations.

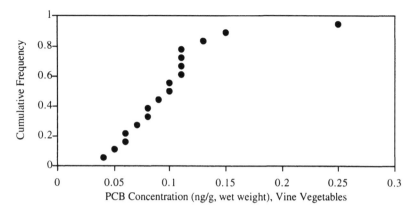

FIGURE 5.6 Empirical distribution of vine produce PCB concentrations.

Several of these are illustrated here using the example of the synthetic dataset of n = 100 drawn at random from a standard normal distribution. The graphical examples presented here were developed using the standard features of the Systat program, which is a statistical analysis program. The Apple Macintosh computer version of this package was used here.

A dot plot, also referred to as a stem and leaf plot, provides an indication of the central tendency and the spread of a data set. There are at least two types of dot plots. Dot plots typically have only one axis (usually horizontal) that has the units of the random variable. While the vertical axis is not given units, the relative number of data points associated with specific values of the random variable is represented by a vertical stacking of dots. One type of dot plot is shown in Fig. 5.8, while another version is shown in Fig. 5.9. Both of these plots indicate that most of the values of this random variable are near zero, with values as

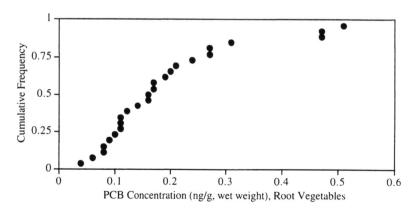

FIGURE 5.7 Empirical distribution of root produce PCB concentrations.

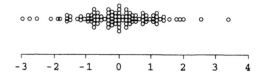

FIGURE 5.8 Dot plot of 100 random samples drawn from a standard normal distribution.

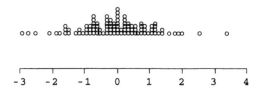

FIGURE 5.9 Alternative version of a dot plot of 100 random samples drawn from a standard normal distribution.

high as approximately +3 and as low as approximately −3. These plots also give an indication regarding the skewness and kurtosis of the data set.

Another convenient graphical representation is the "box and whiskers" plot. A version of this is shown in Fig. 5.10. In the box plot, the median of the data is represented by a line in the middle of the central box. The endpoints of the box represent the interquartile range from the 25th to the 75th percentile. The end points of the whiskers typically represent a 90 or 95% probability range. In the version shown in the figure, some of the extreme data points are also shown, to give a more full indication of the dispersion of data.

It may also be convenient to represent data using a histogram, as is shown in Fig. 5.11. To create a histogram, the user must define the end points of the bins into which the data are to be categorized. The number of data values enclosed within each bin are counted. The number of data points, or the fraction of the total number of data points, are plotted as bars for each of the bins. The histogram provides information regarding the central tendency and shape of the distribution.

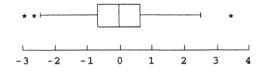

FIGURE 5.10 Box and whiskers plot of 100 random samples drawn from a standard normal distribution.

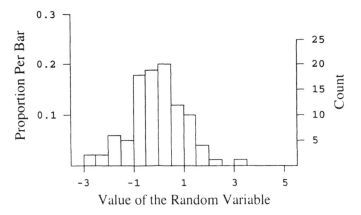

FIGURE 5.11 Histogram of 100 random samples drawn from a standard normal distribution.

5.5. UNCERTAINTY IN SUMMARY STATISTICS

As indicated previously, one can only estimate the values of the mean, variance, and fractiles of a distribution from a data set. The true values of these parameters of a frequency distribution are generally unknown. Thus, there is uncertainty associated with estimates of these quantities.

Uncertainty in the parameters of a distribution, such as a mean or variance, due to finite sample size is represented by a sampling distribution. Sampling distributions can be used to estimate confidence intervals for the parameters of a distribution or to characterize uncertainty due to random sampling error.

5.5.1. Uncertainty in the Mean

There are a variety of methods for characterizing uncertainty in mean values. There are analytical solutions for cases in which the underlying distribution for a data set is normal when the true variance is unknown and when the true variance is known. These analytical solutions may also be useful approximations in some cases where the underlying distribution for the data is not normal. However, it is possible to develop numerical simulations of sampling distributions, using bootstrap simulation. Bootstrap simulation can be used in conjunction with many alternative assumptions regarding the underlying population distribution. All of these methods are described.

5.5.1.1. Uncertainty in the Mean When the True Variance Is Unknown. If we take a random sample of size n from a population that has a mean, μ, and variance, σ^2, then we will only be able to make an estimate, $\bar{x}$, of the true mean and an estimate, s^2, of the true variance. The uncertainty regarding the estimate of the mean can be described, in part, by the sampling variance for the mean:

$$\text{Var}[\bar{x}] = \frac{s^2}{n} \tag{5.21}$$

The standard deviation of a sampling distribution is also referred to as the "standard error." The standard error provides an indication of the precision of estimation of the mean.

The confidence interval on the mean is given by the following expression:

$$\bar{x} \pm c \frac{s}{\sqrt{n}} \tag{5.22}$$

where the mean, standard deviation, and number of data points are obtained from the data set. The value of c is determined based upon the desired confidence level. The confidence level is typically specified as $1 - \alpha$, where α typically assumes values of 0.10, 0.05, or 0.01 for 90, 95, and 99 percent confidence intervals, respectively. Another statement of the confidence interval is the following:

$$Pr\left(-c \leq \frac{\bar{x} - \mu}{s/\sqrt{n}} \leq c\right) = 1 - \alpha \tag{5.23}$$

This equation states the probability that the true value of the mean is enclosed by the confidence interval is equal to the specified confidence level, $1 - \alpha$. The value of c depends on our assumptions regarding our knowledge (or lack thereof) of the true value of the variance and on the shape of the sampling distribution for the mean. See Chap. 9 for an alternate notation.

The type of distribution which applies to uncertainty in the mean depends on the nature of the distribution which underlies the observed data. If the data are observations from a normal distribution, and if the true value of the standard deviation is unknown, then the sampling distribution for the quantity $[(\bar{x} - \mu)/(s/\sqrt{n})]$ is a Student's t-distribution, as introduced in Chap. 4. Thus, one can use the estimate of the standard error for the mean, together with a probability table for the t-distribution, to construct confidence intervals for the mean. The parameter of the Student's t-distribution is the degrees of freedom, which for the uncertainty in the mean is equal to the sample size minus one. Thus, if we have three data points, there are two degrees of freedom. Some typical values of the t-distribution are tabulated in Table 5.2, calculated using the TINV function in

TABLE 5.2 Values from the Student's t-Distribution for Various Degrees of Freedom and Confidence Levels.

Sample Size (n)	Degrees of Freedom (df)	Values of the Student t-distribution for Confidence Level, α		
		$\alpha = 0.10$ (90% CI)	$\alpha = 0.05$ (95% CI)	$\alpha = 0.01$ (99% CI)
2	1	6.314	12.706	65.656
3	2	2.920	4.303	9.925
4	3	2.353	3.182	5.841
5	4	2.132	2.776	4.604
6	5	2.015	2.571	4.032
7	6	1.943	2.447	3.707
8	7	1.895	2.365	3.499
9	8	1.860	2.306	3.355
10	9	1.833	2.262	3.250
11	10	1.812	2.228	3.169
12	11	1.796	2.201	3.106
13	12	1.782	2.179	3.055
14	13	1.711	2.160	3.012
15	14	1.761	2.145	2.977
16	15	1.753	2.131	2.947
17	16	1.746	2.120	2.921
18	17	1.740	2.110	2.898
19	18	1.734	2.101	2.878
20	19	1.729	2.093	2.861
25	24	1.711	2.064	2.797
30	29	1.699	2.045	2.756
40	39	1.685	2.023	2.708
60	58	1.671	2.001	2.662
80	79	1.664	1.990	2.639
100	99	1.660	1.984	2.626
∞	∞	1.645	1.960	2.576

the Microsoft Excel program. Most probability texts provide a more complete set of values.

To illustrate the use of the table of the Student's t-distribution, consider the case for a 95% confidence interval for the mean obtained from the leafy vegetable data set of Chap. 9. The estimated mean value is 0.22 ng PCBs/g (wet weight). The unbiased estimate of the standard deviation is 0.099. There are nine data points. Since the true standard deviation is unknown, we use the Student's t-distribution as the basis for estimating the confidence interval in the mean. A 95% confidence interval corresponds to a confidence level of $\alpha = 0.05$. With nine data points, there are eight degrees of freedom for the sampling distribution of the mean. From the table of the Student's t-distribution, we look up a value of

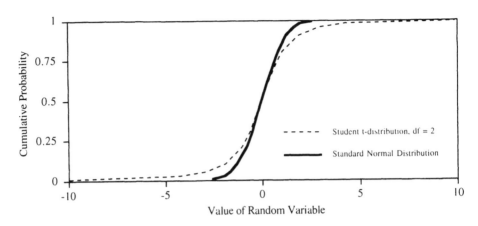

FIGURE 5.12 Comparison of Student's *t*-distribution with two degrees of freedom and standard normal distribution.

$c = 2.365$. Thus, the 95% confidence interval for the mean is (0.14, 0.30). In this case, the uncertainty in the mean for a 95% confidence interval is approximately a factor of two from the lower to the upper confidence bound. This finding indicates that there is a large amount of uncertainty regarding the true value of the mean due to three factors: (1) the large coefficient of variation of the data set; (2) the small sample size; and (3) the fact that the true variance is unknown. If the true value of the variance were known, then the confidence interval on the mean would be narrower. We will return to this later.

As the degrees of freedom increases, the Student's *t*-distribution approaches a normal distribution. A comparison of the Student's *t*-distribution with two degrees of freedom versus the standard normal variate is shown in Fig. 5.12. For small degrees of freedom, the *t*-distribution has much wider tails than does the normal distribution. As the sample size and, hence, degrees of freedom increases, the *t*-distribution converges to a normal distribution. As a practical matter, the normal distribution is typically a reasonable approximation of the sampling distribution of the mean of a normally distributed data set with unknown variance if the number of data points is approximately 30 or more. There is additional discussion of normality approximations, based upon the Central Limit Theorem, in Section 7.1.

5.5.1.2. Uncertainty in the Mean When the Variance Is Known. The sampling distribution for the mean is exactly normal if the variance (or standard deviation) is known and if the data set is from a normal distribution. While this case may appear to be of no practical significance, it is possible that one may have many surrogate data sets for a quantity from which one could obtain a very good estimate of either the standard deviation or coefficient of variation. However, there may be a lack of good site-specific data from which to estimate the mean. For example, the coefficient of variation for ingestion of certain types of foods may be well known, but the average value for a given local population may be poorly specified. Thus, if one has an estimate of the mean based upon limited data, one may be able to calculate a confidence interval assuming that the true standard deviation is approximately known.

As an example, let's suppose that the estimated standard deviation of the leafy vegetable PCB concentration is known to be exactly 0.099. Assuming this, we will calculate the 95% confidence interval for the mean. In this situation, we can use the standard normal variate as the basis for estimating the value of c in Eq. (5.23). For a 95% confidence interval, we are interested in the value of the standard normal associated with the 97.5 percentile, which is 1.96. This value is associated with a two-tailed confidence interval bounded by the 2.5 and 97.5 percentiles. Thus, the confidence interval for the mean is (0.16, 0.28). As expected, this interval is slightly narrower than that for the previously calculated case of unknown variance.

For large enough sample sizes (e.g., >30), for practical purposes the sampling distribution for the mean is itself a normal distribution with a mean value of $\bar{x}$ and a variance of s^2/n. If the sample is drawn from an unknown probability distribution, and if the sample size is large enough, then the sampling distribution for the mean will also be approximately normal. This is true even if the distribution for the data set is skewed. This approximation works well even for relatively small sample sizes for cases in which the population is continuous, unimodal, and symmetric. The approximation may fail when these conditions are not met.

We present three figures to more fully illustrate the confidence interval for the mean of a normal distribution. The first two are examples based upon a standard normal distribution, which has a mean of zero and a variance of one. In Fig. 5.13, we assume the true variance is unknown, but our estimate of the mean and variance are 0 and 1, respectively, for each sample size considered. This is an artificial example, but it serves to highlight the change in the range of the confidence interval due to changes in sample size. In this case, the confidence interval for the mean was calculated based upon the Student's t-distribution. In the second case, shown in Fig. 5.14, we assume the true variance is known and is equal to one. As expected, the confidence intervals for unknown variance are wider than for known variance. However, the difference becomes less pronounced as the sample size increases.

Fig. 5.15 illustrates how the estimates of both the mean and the confidence interval of the mean changes as a function of sample size. This example is for a data set that is normally distributed, with a mean of 1000 and a standard deviation of 500. The figure illustrates that for small sample sizes, the estimate of the mean may be relatively unstable due to random sampling error. Fluctuations in

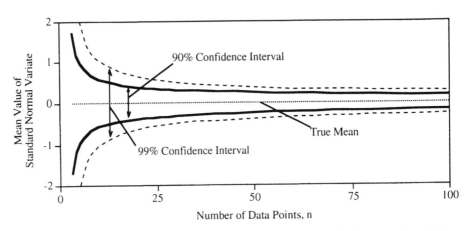

FIGURE 5.13 Confidence interval based upon the Student's t-distribution for the mean of a standard normal variate, $N(0,1)$, assuming the true variance is known.

the estimate of the mean are seen even for relatively large sample sizes of 50 to 100. The confidence interval for the mean based upon small sample sizes is relatively wide compared to the confidence intervals for samples sizes of approximately 30 or more.

5.5.1.3. Bootstrap Simulation of Uncertainty in the Mean. If the underlying population distribution is not normal and if we have only a small sample of data, then the normal distribution may not be an appropriate representation of the sampling distribution for the mean. The approaches to characterizing the sam-

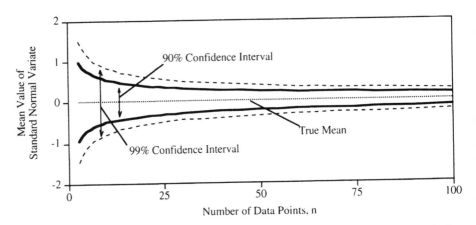

FIGURE 5.14 Confidence interval based upon the normal distribution for the mean of a standard normal variate, $N(0,1)$, assuming the true variance is known.

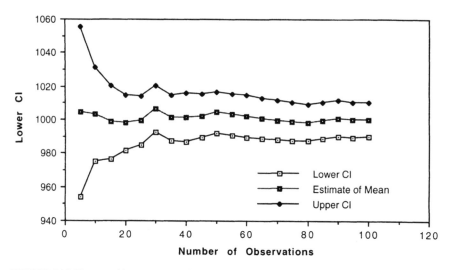

FIGURE 5.15 Upper and lower 95% confidence bounds based on the sampling distribution for the mean.

pling distribution for the mean described in the previous two sections are parametric methods, because they require assumptions regarding a particular parametric family of population distributions (e.g., normal distributions). As an alternative, nonparametric or distribution-free approaches may be used. These approaches do not require assumptions regarding the probability model for the underlying population distribution. However, they also tend to yield wider estimates of confidence intervals than do parametric methods. Alternatively, numerical methods, such as bootstrap simulation, may be used to estimate confidence intervals for the mean or other statistics (Efron and Tibshirani, 1993). Numerical methods such as bootstrap may be based upon either nonparametric or parametric approaches.

The word "bootstrap" can be used to describe a wide variety of specific methods. Here we consider the use of the percentile bootstrap method, known as bootstrap-p, to estimate the confidence interval for the mean of the leafy vegetable data set. A detailed analysis of this data set is given by Frey and Burmaster (1998). Given a data set of sample size n, the general approach in bootstrap simulation is to assume a distribution which describes the quantity of interest, to perform r replications of the data set by randomly drawing, with replacement, n values, and then calculate r values of the statistic of interest. For the first step of assuming a distribution for the data set, there are many options. One approach is to use the actual data set itself and to randomly select, with replacement, the actual values of the data set. This is sometimes referred to as resampling. A second approach is to fit an empirically-based CDF to the data, and to sample from the empirical distribution. For example, one may assume that the data can be described by a cumulative distribution that is piecewise uniform between each data point. Such a distribution has minimum and maximum values constrained by the minimum and maximum values in the data set. This is a referred to here as a fractile distribution, and this type of distribution is briefly described in Chap. 4. A third approach is to assume a parametric distribution, such as normal or lognormal, to represent the data. Each approach will lead to a different estimate of the confidence interval. We will explore each of these approaches using 1500 replications. We will focus on the leafy produce data set. Later, we will summarize results for the vine and root produce data sets.

The results of six different methods for estimating the sampling distributions for the mean of the leafy produce data set are shown in Fig. 5.16. The six approaches provide similar results. If we assume that the underlying data are normally distributed, but that the true variance is unknown, then the Student's *t*-distribution may be used to represent the sampling distribution of the mean. Of the six methods considered, this one yields the widest distribution, although in this example it is not much wider than the others. Alternatively, if the true variance is assumed to be known, then a normal distribution may be used to represent the sampling distribution of the mean (shown as the "Normal distribution" case in Fig. 5.16). This result is nearly identical to the case of a bootstrap simulation in which a normal distribution is assumed to represent the data (shown as the "Bootstrap, Normal Distribution" case). In the bootstrap simulation of the normal distribution, it is assumed that the mean and standard deviation of the

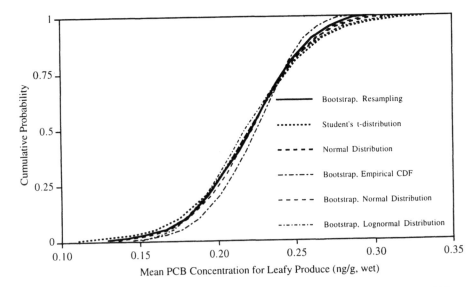

FIGURE 5.16 Estimates of the sampling distribution of the mean for the leafy produce PCB concentration data set.

data set are the true values, and that nine data points are drawn at random from the distribution in each replication. Therefore, the parametric bootstrap simulation is equivalent to assuming a known variance. Thus, it is not surprising that the results of the bootstrap-p approach based upon the normal distribution agree with the results where a normal distribution is imposed on the mean. If alternative distributions are assumed for the data set, then other sampling distributions are obtained. Examples are shown in the cases of resampling, the use of a fractile (empirical) distribution, and the use of a lognormal distribution. This case study demonstrates that it is possible to obtain different sampling distributions for the mean even for a given data set depending on what assumptions the analyst makes regarding the data. In this example, we do not know which one is the "right" answer, if any of these can be considered correct. Thus, judgment is involved in selecting and interpreting a method for estimating the confidence interval or sampling distribution.

It is possible to develop a bootstrap simulation in which the variance of the population distribution is assumed to be unknown. This type of bootstrap is known as Studentized bootstrap, or bootstrap-t (Efron and Tibshirani, 1993). The bootstrap-t approaches require the use of a statistical estimator for the standard error of the statistic for which the sampling distribution is to be evaluated. For this reason, it is a more difficult method to use than bootstrap-p and, in many cases, there may not be an exact solution available for the standard error of the statistic in question (e.g., an arbitrary percentile of a CDF). Thus, we have employed the bootstrap-p approach here.

5.5.2. Uncertainty in the Variance and Standard Deviation

Similar to the discussion for uncertainty in the mean, there are both analytical and numerical methods for characterizing uncertainty in the variance or standard deviation of a data set. We present examples of both of these in the following subsections.

5.5.2.1. Analytical Solution for Uncertainty in Variance. The sampling distribution for the variance is based upon a chi-square distribution if the underlying distribution for the data is normal. For this case, the chi-square distribution can be used to construct confidence intervals on the variance. The parameter of the chi-square distribution is the degrees of freedom. Selected values of the chi-square distribution are given in Table 5.3. These values were calculated using a built-in function in the Microsoft Excel program.

The confidence interval of the variance of a normal population is given by:

$$Pr\left(c_{\alpha/2,n-1} \le \frac{(n-1)s^2}{\sigma^2} \le c_{1-\alpha/2,n-1}\right) = 1 - \alpha \qquad (5.24)$$

TABLE 5.3 Values from the Chi-Square (χ^2) Distribution for Various Degrees of Freedom and Percentiles.[a]

Sample Size (n)	Degrees of Freedom (df)	If X has a χ^2 distribution with df degrees of freedom, this table gives the values of x such that $Pr(X \le x) = p$					
		$p = 0.005$	$p = 0.025$	$p = 0.05$	$p = 0.95$	$p = 0.975$	$p = 0.995$
2	1	0.000	0.001	0.004	3.841	5.024	7.879
3	2	0.010	0.051	0.103	5.991	7.378	10.60
4	3	0.072	0.216	0.352	7.815	9.348	12.84
5	4	0.207	0.484	0.711	9.488	11.14	14.86
6	5	0.412	0.831	1.145	11.07	12.83	16.75
7	6	0.676	1.237	1.635	12.59	14.45	18.55
8	7	0.989	1.690	2.167	14.07	16.01	20.28
9	8	1.344	2.180	2.733	15.51	17.53	21.95
10	9	1.735	2.700	3.325	16.92	19.02	23.59
11	10	2.156	3.247	3.940	18.31	20.48	25.19
12	11	2.603	3.816	4.575	19.68	21.92	26.76
13	12	3.074	4.404	5.226	21.03	23.34	28.30
14	13	3.565	5.009	5.892	22.36	24.74	29.82
15	14	4.075	5.629	6.571	23.68	26.12	31.32
16	15	4.601	6.262	7.261	25.00	27.49	32.80
17	16	5.142	6.908	7.962	26.30	28.85	34.27
18	17	5.697	7.564	8.672	27.59	30.19	35.72
19	18	6.265	8.231	9.390	28.87	31.53	37.16
20	19	6.844	8.907	10.12	30.14	32.85	38.58
25	24	9.886	12.40	13.85	36.42	39.36	45.56
30	29	13.12	16.05	17.71	42.56	45.72	52.34
40	39	20.00	23.65	25.70	54.57	58.12	65.48
60	59	34.77	39.66	42.34	77.93	82.12	90.72
80	79	50.38	56.31	59.52	100.7	105.5	115.1
100	99	66.51	73.36	77.05	123.2	128.4	139.0

[a]Values for this table were calculated using the CHINV function in the Microsoft Excel 5.0 program

It is perhaps easiest to illustrate this with an example. For the case of the leafy produce data set, the estimated standard deviation is 0.099 ng/g (wet basis) and, therefore, the estimated variance is 9.7×10^{-3} (ng/g)^2. If we assume for the moment that the distribution underlying the PCB concentration in leafy produce is normal, then we can assume that the following quantity is distributed as a chi-square distribution:

$$\frac{(n-1)s^2}{\sigma^2} \sim \chi^2_{n-1} \tag{5.25}$$

The degrees of freedom are $n - 1$, where n is the number of data points. For a given confidence level, α, we can look up or calculate values of the chi-square distribution. For example, suppose that $\alpha = 0.05$, which translates to a 95% confidence interval. Such an interval would be bounded by the 0.025 and 0.975 fractiles of a CDF. Since we have $n = 9$, we can go to Table 5.3 and look up the values of the chi-square distribution with eight degrees of freedom for the selected values of the CDF. We find that these numbers are 2.180 and 17.53, respectively, for the 0.025 and 0.975 fractiles. These two numbers can be represented as $c_{0.025,8}$ and $c_{0.975,8}$, using the notation of Eq. (5.24). Given these values plus our estimate of the variance from the data set, we can estimate the interval such that we have 95% confidence that the true variance is enclosed as follows:

$$Pr\left(\frac{(n-1)s^2}{c_{1-\alpha/2,n-1}} \leq \sigma^2 \leq \frac{(n-1)s^2}{c_{\alpha/2,n-1}}\right) = 1 - \alpha \qquad (5.26)$$

and for our example becomes:

$$Pr\left(\frac{(9-1)(9.86 \times 10^{-3})}{17.53} \leq \sigma^2 \leq \frac{(9-1)(9.86 \times 10^{-3})}{2.180}\right) = 1 - 0.05$$

$$Pr(0.0045 \leq \sigma^2 \leq 0.036) = 0.95$$

This result indicates that with 95% confidence the true variance is enclosed within the range from 0.0045 to 0.036 $(ng/g)^2$. This translates into a 95% confidence interval for the standard deviation of 0.067 to 0.191 ng/g. Of course, we may question the validity of this finding if the underlying population is not normally distributed. Later we will argue that the normal distribution is not the most appropriate selection to represent this data set. For now, we will explore the chi-square distribution via some sensitivity analyses, and then return to the evaluation of the confidence interval on the standard deviation of the leafy produce data set using bootstrap simulation methods for comparison with the solution obtained using the chi-square distribution.

As shown in Fig. 5.17, the confidence intervals are asymmetric with respect to the estimated variance. The example shown is for a population that has a normal distribution with a mean of zero and a variance of one. If for a given sample size an estimate of the variance of one is obtained, the resulting 90 and 99% confidence intervals on the variance are shown. The chi-square distribution tends to be positively skewed, especially for small sample sizes. This is because variance is a non-negative quantity. Furthermore, the confidence intervals are quite wide for small sample sizes. For sample sizes even as large as 100, there is significant uncertainty regarding the variance due to finite sample size.

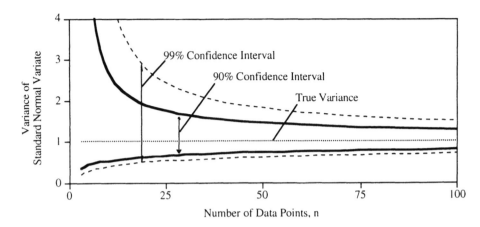

FIGURE 5.17 Confidence intervals on the variance as a function of sample size for a standard normal distribution.

To illustrate both the random sampling error in estimates of the variance and the confidence interval for the variance, we performed a numerical simulation. The results are shown in Fig. 5.18. This example is based upon numerical simulation of a normal distribution with a mean of 1000 and a standard deviation of 50. Using random Monte Carlo sampling (described in Chap. 7) simulations for different sample sizes from $n = 3$ to $n = 100$ were performed. For each

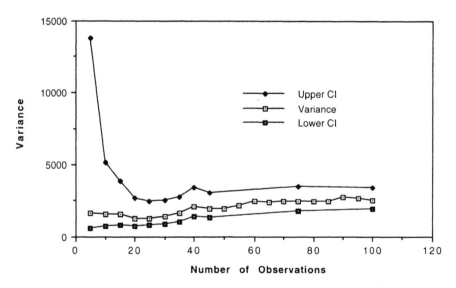

FIGURE 5.18 Confidence intervals on the variance for samples from a normal distribution with a mean of 1000 and Variance of 2500.

simulation, an estimate of the variance was calculated. Furthermore, for each sample size, a confidence interval for the variance was calculated using the procedure involving the chi-square distribution. The results indicate there can be substantial fluctuation in the value of the estimated variance due to random sampling error, but as the sample size increases one may expect the estimated variance to more accurately converge toward the true value. The confidence interval is extremely large for small sample sizes, but is still relatively large even for 100 samples.

5.5.2.2. Bootstrap Simulation of Uncertainty in Standard Deviation. The chi-square distribution is not universally applicable as a representation of the sampling distribution for the variance for any type of underlying probability distribution. As shown in Fig. 5.19, the shape of the sampling distribution depends upon the underlying probability distribution for the population. This example is for the leafy produce data set. The cases shown include resampling of the original data set, specification of an empirical distribution for the population using a fractile distribution, and specification of parametric distributions for the data set using normal and lognormal distributions. Furthermore, a sampling distribution for the standard deviation based upon the chi-square distribution for the variance is also shown. The case studies based most directly upon the data set, which include both the resampling and empirical CDF cases, yield the narrowest confidence intervals. This is because both case studies do not extrapolate beyond the

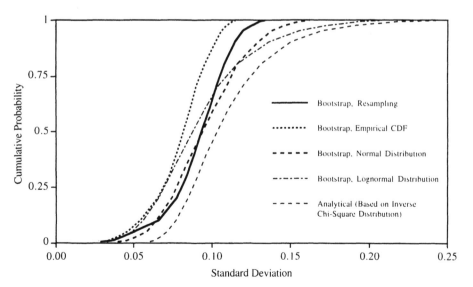

FIGURE 5.19 Estimates of the sampling distribution of standard deviation for the leafy produce PCB concentration data set.

minimum and maximum values of the data sets. The two parametric cases, and the analytical solution based upon the chi-square distribution, assume that the underlying population is unbounded. For the lognormal case, it is assumed that the population is non-negative and unbounded for positive values. Thus, as expected, these cases give wider confidence intervals for the variance than the resampling or empirical CDF cases.

5.5.3. Uncertainty in Skewness

Sampling distributions can be constructed for any statistic which describes a data set. For example, Bowman and Shenton (1986) present isopleths and scatterplots illustrating the random variation in skewness and kurtosis due to sampling error. It is important to recognize that estimates of moments or distribution parameters may be subject to wide confidence intervals, especially for small data sets or when the variation within the data set is very large. Thus, it is sometimes difficult to rely on data analysis alone as the basis for fitting probability distributions to data sets.

As an example of uncertainty in skewness, we consider the leafy produce PCB concentration data set of Chap. 9. For this data set, we have previously calculated an estimate of the skewness, which was found to be –0.58 (as given in Table 5.1). Using the bootstrap simulation methods previously described, estimates of uncertainty regarding the skewness were developed and are displayed in Fig. 5.20. Of the four cases shown, the resampling and empirical CDF approaches provide nearly identical results. These two cases have a central value similar to the point estimate, and indicate that the 95% confidence interval ranges from approximately –1.4 to +0.4. If the population is assumed to be normally distributed, then the 95% confidence interval for the skewness is –1.1 to +1.1. Thus, the range is approximately symmetric about zero, which is the expected value for the skewness of a normal population. If the population is assumed to be lognormal, then the 95% confidence interval for the skewness is –0.5 to +1.6. The sampling distribution based upon the lognormal indicates that the expected skewness is positive. However, with only nine data points, it is possible to obtain sets of samples from a lognormal distribution which may have negative skewness. This last insight highlights the importance of not placing too much confidence in point estimates of statistics based upon small sample sizes. Although the leafy data set has a negative skewness, it is possible that this data set is consistent with a population that is positively skewed.

5.5.4. Uncertainty in Kurtosis

Similar to the case studies for the uncertainty in the skewness presented in the previous section, we consider here uncertainty regarding the kurtosis of a data

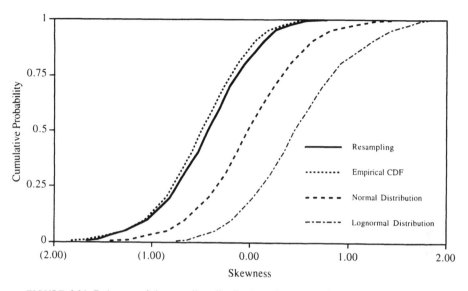

FIGURE 5.20 Estimates of the sampling distribution of skewness for the leafy produce PCB concentration data set.

set. As before, four cases are considered based upon the numerical method of bootstrap simulation for the leafy produce PCB concentration data set of Chap. 9. The results are displayed in Fig. 5.21. It appears that the sampling distribution of the kurtosis is less sensitive to the underlying assumptions regarding the distribution of the data set than was the case for the sampling distributions of the skewness. In all four cases considered, the 95% confidence interval for the kurtosis ranges from approximately 1.0 to 4.0. This range of values implies that a number of parametric distributions may provide plausible fits to the data set, based upon inspection of the β_1-β_2 plane in Fig. 5.2. The uniform, beta, normal, gamma, and lognormal distributions can accommodate various combinations of the values of skewness and kurtosis that have been displayed within the sampling distributions for both of these statistics for the example of the leafy produce PCB concentration data set.

5.5.5. Multivariate Distributions for Uncertainty in Statistics

In the previous four sections, we have provided examples of the marginal sampling distributions of the mean, variance (or standard deviation), skewness, and kurtosis. In this section, we look at some of the relationships between these statistics. Specifically, we consider the bivariate distributions for the mean and standard deviation and for the skewness and kurtosis, and describe some of the

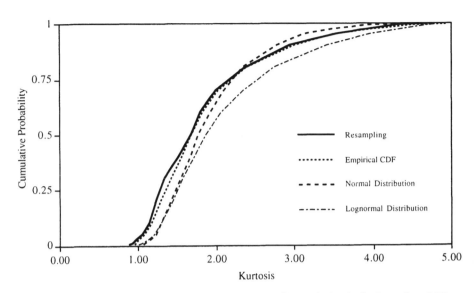

FIGURE 5.21 Estimates of the sampling distribution of kurtosis for the leafy produce PCB concentration data set.

insights that these bivariate distributions afford regarding the selection of probability models for representing data sets.

For some types of distributions, it can be shown that the sampling distributions of the mean and variance are statistically independent. For example, this is true of the normal distribution (Laha and Rohatgi, 1979). Furthermore, the mean and variance of the logarithm of a data set are independent if the data are lognormally distributed. Similarly, the geometric mean and geometric standard deviation of a lognormally distributed data set are also independent. However, in other cases, there may be dependence between the sampling distributions. This is true, for example, for the arithmetic mean and variance of lognormally distributed data, or for the parameters of many other types of distributions. Examples of the latter include the shape parameters of the beta distribution, and the shape and scale parameters of the gamma and Weibull distributions. Frey and Rhodes (1998) explore the dependencies between sampling distributions for various types of probability models in more detail.

Fig. 5.22 illustrates some of the different bivariate sampling distributions that may be observed for the mean and standard deviation. The figure is based upon bootstrap simulations for the leafy produce PCB concentration data set of Chap. 9, as detailed in Frey and Burmaster (1998). Fig. 5.22(a) is based upon resampling of the actual data set. Because the data set is negatively skewed, there is a negative correlation between the standard deviation and the mean. However,

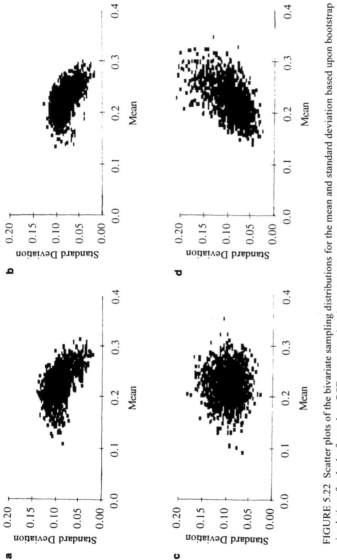

FIGURE 5.22 Scatter plots of the bivariate sampling distributions for the mean and standard deviation based upon bootstrap simulations for the leafy produce PCB concentration data set: (a) Resampling; (b) Empirical (fractile) distribution; (c) Normal distribution; (d) Lognormal distribution.

it appears that the dependence between the standard deviation and the mean is nonlinear and perhaps even non-monotonic. A similar result is obtained if an empirical distribution, bounded by the maximum and minimum values of the data set, is assumed, as shown in Fig. 5.22(b). For a normal distribution, there is no dependence between the standard deviation and the mean, as shown in Fig. 5.22(c). For a lognormal distribution, there is a positive correlation between the arithmetic standard deviation and the arithmetic mean, as shown in Fig. 5.22(d).

Because we have only nine data points for the leafy produce PCB concentration data set, we cannot with certainty reject the notion that the population of leafy produce PCB concentrations may be lognormally distributed. In fact, there is substantial overlap between the bivariate distribution for the resampled data and for the assumed lognormal population distribution.

In Fig. 5.23, the bivariate sampling distributions for the skewness and kurtosis are displayed. Similar to the cases for the bivariate distributions for the mean and standard deviation, there are similarities for the cases based upon resampling of the original data and use of an empirical distribution bounded by the minimum and maximum values in the data set. Both of these cases indicate that it is more likely that the data will be negatively than positively skewed, and that as the magnitude of the skewness increases, the kurtosis will also increase. If the data were obtained from a normal distribution, then random sampling error for a random draw of nine data points would lead to equal likelihood of observing positively or negatively skewed data sets. If the data were obtained from a lognormal distribution, it would be more likely that one would observe a positively than a negatively skewed data set. However, it is clear that it is possible that the leafy produce data set, which has a skewness of -0.58 and a kurtosis of 2.2, could have been obtained from a parent population which is lognormally distributed.

The inspection of bivariate sampling distributions is useful in that it allows us to identify whether the combinations of values we estimate for a given data set could plausibly have been obtained from different types of distributions. The bivariate distributions for the skewness and kurtosis clearly indicate that there are combinations of values of skewness and kurtosis which are impossible either in general (as shown in Fig. 5.2) or for some types of population distributions. Furthermore, these graphs illustrate that with small sample sizes there can be substantial uncertainty in the estimates of these statistics.

5.5.6. Small Data Sets

Concern is often expressed with what to do with a small number of data points. If the data are representative of the quantity of interest, then the concern may be focused on how to quantify the random sampling error in the data set. If the data are nonrepresentative, then statistical analysis will provide an insufficient basis for quantifying either variability or uncertainty in the data.

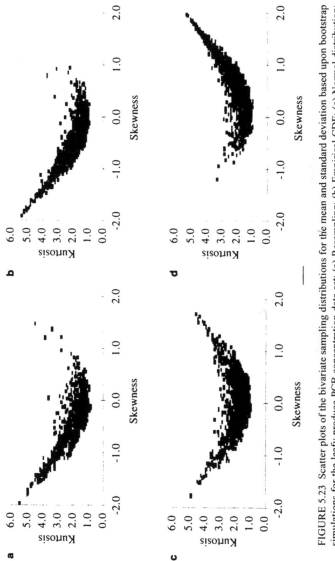

FIGURE 5.23 Scatter plots of the bivariate sampling distributions for the mean and standard deviation based upon bootstrap simulations for the leafy produce PCB concentration data set: (a) Resampling; (b) Empirical CDF; (c) Normal distribution; (d) Lognormal distribution.

Assuming we have a small number of data points that are a random sample of a quantity representative of what is needed as input to an assessment, then classical statistical techniques can be used to calculate the central moments of the data set and to estimate the parameters of distributions. For example, even with two data points, it is possible to estimate the mean and variance of a normal or lognormal distribution. Of course, the uncertainty in the estimates will be large. However, the uncertainty can be quantified based upon sampling distributions for the parameters. For example, we have reviewed the use of the Student's *t*-distribution and the chi-square distribution as the basis for estimating confidence intervals for the mean and variance of a normal distribution.

In Table 5.4, the 95% confidence intervals for the mean and variance are presented for a data set with *n* data points for which the estimated mean is 0.0 and variance is 1.0. With only two data points, there is a great deal of uncertainty in the estimates of both the mean and the variance. A third data point substantially reduces this uncertainty. However, even with ten data points, there is still a large amount of uncertainty regarding the true values of the parameters. Thus, random sampling error can contribute significantly to the total uncertainty regarding the data set. Additional sources of uncertainty may include measurement error and nonrepresentativeness of the data.

5.5.7. Summary of Uncertainty in Statistics for the PCB Concentration Data Sets

A summary of uncertainty in the mean, standard deviation, skewness, and kurtosis for three data sets is given in Table 5.5. The uncertainty is summarized

TABLE 5.4 Examples of Confidence Intervals for the Parameters of Distributions Based upon Small Data Sets: Mean and Variance of $N(0,1)$

	95% Confidence Interval for the Mean Based upon Student's *t*-Distribution		95% Confidence Interval for the Variance based upon the Inverse Chi-Square Distribution	
n	LCB	UCB	LCB	UCB
2	−8.98	8.98	0.199	1,018
3	−2.48	2.48	0.271	39.5
4	−1.59	1.59	0.321	13.9
5	−1.24	1.24	0.359	8.26
6	−1.05	1.05	0.390	6.02
7	−0.925	0.925	0.415	4.85
8	−0.836	0.836	0.437	4.14
9	−0.769	0.769	0.456	3.67
10	−0.715	0.715	0.473	3.33

TABLE 5.5 Summary of 95% Confidence Intervals for the Mean, Standard Deviation, Skewness, and Kurtosis for PCB Concentrations in Leafy, Vine, and Root Produce: Comparison of Analytical and Boostrap Methods.

		Data Set					
		Leafy Produce 95% Confidence Interval		Vine Produce 95% Confidence Interval		Root Produce 95% Confidence Interval	
Statistic	Method	Lower	Upper	Lower	Upper	Lower	Upper
Mean	Student's t-Distribution	0.145	0.297	0.0751	0.125	0.140	0.247
	Normal Distribution	0.156	0.286	0.0770	0.123	0.143	0.244
	Bootstrap - Resampling	0.153	0.280	0.0806	0.125	0.146	0.246
	Bootstrap - Empirical	0.167	0.270	0.0802	0.117	0.144	0.239
	Bootstrap - Normal	0.158	0.283	0.0759	0.123	0.144	0.245
	Bootstrap - Lognormal	0.163	0.293	0.0783	0.125	0.148	0.249
Standard Deviation	Inverse Chi-Square Distribution	0.067	0.190	0.036	0.074	0.101	0.180
	Bootstrap - Resampling	0.044	0.125	0.022	0.070	0.073	0.164
	Bootstrap - Empirical	0.039	0.109	0.020	0.055	0.067	0.150
	Bootstrap - Normal	0.052	0.146	0.032	0.065	0.092	0.167
	Bootstrap - Lognormal	0.043	0.173	0.026	0.077	0.070	0.220
Skewness	Bootstrap - Resampling	-1.43	0.41	-0.51	2.11	0.42	1.88
	Bootstrap - Empirical	-1.48	0.34	-0.55	1.94	0.34	1.98
	Bootstrap - Normal	-1.07	1.05	-0.89	0.86	-0.88	0.84
	Bootstrap - Lognormal	-0.51	1.60	-0.13	2.21	0.21	2.90
Kurtosis	Bootstrap - Resampling	0.98	3.96	1.50	7.94	1.66	6.94
	Bootstrap - Empirical	1.01	4.00	1.40	7.29	1.74	7.42
	Bootstrap - Normal	1.16	3.49	1.48	4.13	1.66	4.51
	Bootstrap - Lognormal	1.13	4.36	1.52	8.02	1.77	12.40

based upon the 95% confidence interval for each statistic. Three data sets are included. These are the leafy, vine, and root produce PCB concentration data sets of Chap. 9. In all cases, confidence intervals were developed using four different bootstrap simulations based upon 1500 replications of an assumed underlying distribution. These assumed distributions include resampling of the actual data values, an empirical CDF bounded by the minimum and maximum data values, and normal and lognormal distributions with the same mean and standard deviation as the original data set. In the case of the mean and standard deviation, other methods presented in this chapter were also employed. These include the use of the Student's t-distribution and normal distribution for the sampling distribution of the mean, and the chi-square distribution as the basis for the sampling distribution of the variance. Thus, six methods were used to estimate the confidence interval for the mean, five methods were used to estimate the confidence interval of the standard deviation, and four methods were used for both the skewness and kurtosis. More detail regarding the analysis of variability and uncertainty due to random sampling error for these data sets are given in Frey (1997), Frey (1998), and Frey and Burmaster (1998).

The confidence intervals for the means obtained from the six different approaches were approximately similar for each of the three data sets. These results indicate that the estimate of uncertainty in the mean may not be highly sensitive to the choice of method in these examples. However, one should not infer that similar results would be obtained in all cases. For smaller data sets, or data sets that are highly skewed, one may obtain more significant differences among these approaches.

The results for the confidence intervals of the standard deviations of the three data sets appear to be more highly dependent upon the choice of method and assumptions than was the case for the means. For example, the upper 95% confidence bound for the standard deviation of the leafy produce data set ranges from 0.109 to 0.180 depending upon which method is used. It is typically, although not always, the case that confidence intervals based upon assumptions of unbounded distributions (e.g., normal, lognormal) yield larger upper confidence bounds than for intervals based upon assumptions of bounded distributions (e.g., resampling, bounded empirical CDF).

For all three data sets, the confidence interval for the skewness appears to depend most strongly upon assumptions regarding the underlying probability distribution that generated the data. When few assumption are made beyond the information contained in the data, such as for the resampling and empirical CDF cases, the confidence intervals are similar to each other and their central tendency reflects the best estimate of the skewness of the data obtained from the method of moments (see Table 5.1). When stronger assumptions are made regarding the underlying probability distribution, the confidence interval shifts. For example, if the underlying distribution is assumed to be normal, then the confidence interval is approximately symmetric with respect to zero. If the underlying distribution is assumed to be lognormal, then the confidence interval for the skewness is shifted toward positive values. The width of the confidence intervals is affected by both the sample size and the coefficient of variation. For example, even though the root produce data set has the most data points, it has the widest confidence interval for the lognormal case because it has the largest coefficient of variation.

The confidence intervals for kurtosis agree most closely when the underlying assumptions regarding the population distribution are similar. For example, when finite bounds are assumed, as in the case of the resampling and empirical bootstrap approaches, the upper and lower bounds of the 95% confidence interval are similar for all three data sets. In the case of the leafy produce data set, all four approaches yield similar results, indicating that with the relatively low magnitude of skewness in the data set, many distributions with low skewness and similar "peakedness" would reasonably represent the data. The lower confidence bound is similar for all four approaches for all three of the data sets. However, for the vine and root data, the upper bound is sensitive to whether the underlying distribution is assumed to be normal or lognormal. The lognormal cases give substantially wider confidence bounds. The lognormal is a more appropriate choice for the vine and root data than is the normal, since the coefficient of variation is large, as previously discussed, and the skewness is highly likely to be nonnegative. Thus, the confidence interval for kurtosis obtained based upon the normal distribution may be artificially too narrow, because in this case we have imposed characteristics upon the data that prevent larger values of kurtosis.

It is hoped this review will give you appreciation for the potentially large sources of uncertainty in summary statistics that can be attributed to random sampling error. Of course, there may be additional sources of error which can sometimes be far more important. For example, nonrepresentative data may be biased and misrepresent the variation, skewness, and kurtosis of the true quantity of interest. In addition, data may embody a combination of true variability and measurement error. When measurement errors are relatively small, such as plus or minus ten percent, random sampling error may in fact be the dominant source of uncertainty for representative data and small sample sizes. For example, the measurement error in the leafy produce data is approximately plus or minus 10%. However, since the mean is uncertain by a factor of two, the standard deviation is uncertain by a factor of three, and the skewness and kurtosis are uncertain by relatively wide ranges that allow for the selection of many plausible parametric probability distribution models, random sampling error appears to be the major source of uncertainty in this case. Even for the root produce data, for which there are 25 data points, random errors are significantly larger than plus or minus 10%.

5.6. PROBABILITY DISTRIBUTION MODELS

A probability distribution model is a description of the probabilities of all possible sets of outcomes in a range space. A probability model is typically represented mathematically as a probability distribution in the form of either a PDF or CDF for a continuous random variable, as described in Chap. 4. There are a number of common functional forms for probability distributions which are used in exposure assessments. The theoretical basis for using a particular probability model is described further in Chap. 4. Alternatively, one may develop an "empirical" distribution to represent a data set.

Probability distribution models are described by their parameters. The power of using probability distributions is that data sets containing potentially large numbers of samples can be described by a probability model that is defined by typically one to four parameters. For example, a normal distribution is fully specified if its mean and variance are known. Small (1990) discusses six characteristics of estimators for the parameters of distributions. These are: consistency, lack of bias, efficiency, sufficiency, robustness, and practicality. A consistent estimator converges to the "true" value of the parameter as the number of samples increases. An unbiased estimator yields an average value of the parameter estimate that is equal to that of the population value. An efficient estimator has minimal variance in the sampling distribution of the estimate. An estimator that makes maximum use of information contained in a data set is said to be sufficient. A robust estimator is one that works well even if there are departures from the assumed underlying distribution. Finally, a practical estimator is one that satisfies the needs for the preceding five characteristics while remaining computationally efficient.

5.6.1. Method of Matching Moments Estimates of Distribution Parameters

Two of the most common approaches to estimating the parameters of a distribution are the method of matching moments and maximum likelihood methods. In the method of matching moments (referred to also as the "method of moments"), the parameters of a probability distribution model are selected so that the moments of the model match the moments of the data set. The number of moments that are used in this process corresponds to the number of parameters to be estimated. This approach is usually the most straightforward to implement. Therefore, it typically satisfies the criteria of practicality. However, it may not fully satisfy the other five criteria.

5.6.2. Maximum Likelihood Estimates of Distribution Parameters

The maximum likelihood method involves selecting values of the distribution parameters that are most likely to yield the observed data set. To do this, a likelihood function is defined. For a continuous random variable for which independent samples have been obtained, the likelihood function is:

$$L(\theta_1, \theta_2, \ldots, \theta_k) = \prod_{i=1}^{n} f(x_i | \theta_1, \theta_2, \ldots, \theta_k) \qquad (5.27)$$

where the likelihood function, L, is evaluated based upon the product of the probability density function evaluated for each sample. The parameters of the probability model, θ_k, are selected so as to maximize the value of the likelihood function. Thus, this is an optimization problem in which the first derivative of the likelihood function with respect to each of the distribution parameters is set to zero. The set of first derivative equations is then solved to obtain the values of the distribution parameters. For small sample sizes, the maximum likelihood estimates do not always yield minimum variance or unbiased estimates (Holland and Fitz-Simons, 1982). However, for larger sample sizes, the maximum likelihood method tends to better satisfy the first five criteria for statistical estimators. However, it is often much more difficult to implement than the method of moments (Small, 1990). Johnson (1980) points out that analysts should not feel the method of maximum likelihood is always the ideal.

5.6.3. Examples of the Use of Statistical Estimators

In some cases, the method of moments and the maximum likelihood method yield the same estimators. For example, the arithmetic average of a data set is the maximum likelihood estimate of the mean of a normal distribution, and the biased estimator for the standard deviation is the maximum likelihood estimator of the standard deviation of a normal distribution. In other situations, the method of moments and maximum likelihood estimates may differ. We consider an example of this for the gamma distribution. Furthermore, in other cases, there may not be a convenient method of moments solution for distribution parameters, and maximum likelihood may be the best or at least a practical alternative. This is the case, for example, for the Weibull distribution.

We begin with an example of the application of maximum likelihood to the estimation of the parameters of a gamma distribution. The probability of one random sample from a gamma distribution (see also as Equ. 4-11) is given by:

$$p(x_i | \alpha, \beta) = f(x_i) = \frac{\beta^{-\alpha} x_i^{\alpha-1} \exp\left(-\dfrac{x_i}{\beta}\right)}{\Gamma(\alpha)} \qquad (5.28)$$

This is simply the probability density function evaluated for an individual data point. The likelihood function is defined in terms of the product of the probabilities of each random sample in the data set. By taking a logarithmic transform of the data, the loglikelihood function can be written in terms of the sum of the logarithms of the probabilities of each sample. Thus, the log-likelihood function for the gamma distribution is given by:

$$J_{\text{gamma}}(\alpha, \beta) = -n\{\alpha \ln(\beta) + \ln[\Gamma(\alpha)]\} + \sum_{i=1}^{n} \left\{ (\alpha - 1)\ln(x_i) - \frac{x_i}{\beta} \right\} \quad (5.29)$$

A simulated data set of 50 values from a gamma distribution, with parameters $\alpha = 1$ and $\beta = 1$, was used to illustrate the application of the loglikelihood function for the gamma distribution. It is possible to do this in the Microsoft Excel program , because a function for the natural logarithm of the gamma function is available. The Solver function in the Excel program was used to iterate on values of the parameters in search of a maximum for the loglikelihood function. The maximum loglikelihood was found to be −53.64 and the maximum likelihood estimates of the parameters were found to be $\alpha = 0.84$ and $\beta = 1.29$. From the method of moments, the parameters were estimated to be $\alpha = 1.16$ and $\beta = 0.94$ based upon the estimated mean and (biased) standard deviation as follows:

$$\hat{\alpha} = \left(\frac{\bar{x}}{s} \right)^2 \quad (5.30)$$

$$\hat{\beta} = \frac{\bar{x}}{s^2} \quad (5.31)$$

These results indicate that the method of moments estimates do not equal the maximum likelihood estimates. The results also indicate that with small sample sizes, there is sampling error in the estimates of the parameters. As the sample size increases, the two methods appear to agree more closely. For example, for a simulated data set of $n = 500$ drawn from a gamma distribution with parameters $\alpha = 1$ and $\beta = 1$, the method of moments estimates for the parameters were $\alpha = 0.87$ and $\beta = 1.06$ and the maximum likelihood estimates were $\alpha = 0.92$ and $\beta = 1.00$.

For the Weibull distribution, the probability of one random sample is given by:

$$p(x_i \mid k, c) = f(x_i) = \frac{k}{c} \left(\frac{x_i}{c} \right)^{k-1} \exp\left[-\left(\frac{x_i}{c} \right)^k \right] \quad (5.32)$$

and the loglikelihood function is given by:

$$J_{\text{Weibull}}(k, c) = n \, \ln\left(\frac{k}{c}\right) + \sum_{i=1}^{n} \left\{ (k - 1)\ln\left(\frac{x_i}{c}\right) - \left(\frac{x_i}{c}\right)^k \right\} \qquad (5.33)$$

The Weibull distribution is an example for which the maximum likelihood esti-mator may be more convenient than the method of matching moments. There is no closed form solution for estimating the parameters of the Weibull distribu-tion from the mean and standard deviation of a data set, because the mean and variance depend upon the gamma function and use arguments that include the distribution parameters. Thus, a numerical solution method is required to use the method of matching moments. The maximum likelihood approach is relatively easy to implement in a spreadsheet such as the Excel program. We applied the maximum likelihood method to two data sets, both obtained from numerical simu-lation of random samples from a Weibull distribution with $k = 2$ and $c = 1$. For the simulated data set of $n = 50$, the maximum likelihood parameter estimates were $k = 2.16$ and $c = 0.97$. For the simulated data set of $n = 500$, the estimates were $k = 2.02$ and $c = 0.98$. These results indicate that sampling error in the estimates of the parameters is reduced as the sample size increases. An alternative approach to estimating the parameters of the Weibull distribution is considered in the next section on probability plotting.

Estimators for various types of distributions are given in many standard texts (e.g., Hahn and Shapiro, 1967; Morgan and Henrion, 1990). Other methods for fitting distributions are discussed in the next section.

5.7. PROBABILITY PLOTS

If you have a data set to which you wish to ascribe some particular type of probability distribution, then you can use the sample values to estimate the pa-rameters of the distribution using the method of moments or the maximum like-lihood method. However, while it may be possible to obtain what appear to be reasonable values for these parameters (e.g., mean, variance, shape, scale, etc.), it may turn out that the selected probability model (e.g., normal, lognormal, Weibull, gamma, beta, etc.) does not offer a good representation of the entire data set. Thus, other methods should be used to make inferences about the "good-ness-of-fit" of a probability model to a data set. Graphical techniques can be used to identify inadequacies of a particular probability model. For example, if the focus of interest is the upper tail of a probability distribution, then a probability model, which offers an excellent fit to the central ranges of the data but a poor fit to the upper tail, might be deemed inadequate for the purposes of a particular assessment.

In this section and the next, we consider two approaches for testing distributional assumptions based upon available data. In this section, we discuss graphical techniques for comparing distributional assumptions with data. These techniques are known as probability plotting. In the next section, we discuss the goodness-of-fit tests further.

Probability plotting is a subjective technique for evaluating whether data contradict an assumed probability distribution. It is a relatively easy and intuitive approach. Statistical tests tend to provide a more objective measure of how well a distribution function represents the data. However, with small data sets statistical tests may often have low "power;" they may fail to reject a model that is in fact a poor fit to the data. Thus, even if you use statistical tests, you should consider also visualizing the data and the model to make a subjective evaluation of the goodness-of-fit. Furthermore, with very large data sets, goodness-of-fit tests may reject fits which may in fact be adequate for a given purpose. For example, if you have a data set with 2000 data points, then statistical tests may indicate that a parametric distribution, such as a lognormal, should be rejected as an inappropriate fit to the data. However, the lognormal distribution may do an adequate job at representing the aspects of the data set of most interest to you (e.g., the central tendency, the tails) within the data quality objectives of your analysis.

Probability plots are advantageous in situations in which one wishes to obtain a good fit for a particular portion of the distribution, such as the upper tail, and to retain a reasonable fit for the remainder of the distribution. Probability plots are also useful in cases involving censored or truncated data sets. For example, it is common to have a data set of measurements for which some portion of the values are not reported because they are below the detection limit of the measurement technique.

5.7.1. Empirical Basis for Selecting Probability Models

The selection of an appropriate distribution model should start with a consideration of the underlying phenomena which generated the data. These phenomena were reviewed in Chap. 4. Thus, you should not arbitrarily select distributional models and pick the one with the best fit to the data. You should narrow the set of possible models to those that are plausible, based on theoretical considerations, and then select the distribution which offers the best combination of theoretical basis and goodness-of-fit to the data. Moreover, you need not select from the "named" set of distributions. It is appropriate, for example, to develop arbitrary empirically-based distributional models if they more realistically represent a data set. The stringency of the requirements for fitting distributions to data will depend on the data quality objectives for your particular exposure assessment. Thus, distributions developed for a first-pass screening analysis of uncertainty and variability may have more relaxed criteria than those developed for a final rule-making or a site-specific analysis to identify required cleanup actions. For example, for a preliminary analysis, a reasonable estimate of the mean and variance may be sufficient for the purposes of identifying key contributors to uncertainty and focusing data collection efforts. However, for a situation in which exposures to high end individuals (e.g., those above the 90th percentile of the population) must be calculated, more stringent requirements may exist regarding correctly specifying the shape of the distributions and especially the tails.

When ambiguity exists regarding what probability model to select, there are some statistics-based methods for making such decisions. Such methods are often based upon criteria related to the moments of the data set. Many types of probability models are actually special cases of more general families of probability models. Some of these families include Gram–Charlier and Edgeworth curves, Pearson curves, Burr (Types III and XII) curves, and Johnson curves (Johnson, 1980). Examples of the Pearson and Johnson systems were reviewed in Section 5.2. These various families yield special cases with a variety of characteristics. For example, Gram-Charlier curves include bimodal distributions. Pearson curves include well-known probability models, such as exponential, uniform, beta, gamma, Student's t, F, and normal distributions. Burr curves include the Weibull distribution. Johnson curves include bimodal distributions.

In many texts (e.g., Hahn and Shapiro, 1967), charts are provided in which different probability models are distinguished based upon their skewness and kurtosis. We have presented examples of these in Figs. 5.2 and 5.3. Thus, by estimating the skewness or kurtosis of a dataset it is possible to identify candidate probability models that may provide a good fit to the data. For example, a normal distribution always has a skewness of zero and a kurtosis of three. Thus, a normal distribution would not be appropriate to represent data with a skewness of four and kurtosis of nine. In this latter case, an exponential distribution would be appropriate.

5.7.2. Methods for Probability Plotting

The probability plotting technique is relatively straightforward to implement, especially when compared to other parameter estimation techniques such as maximum likelihood. Historically, this method involves plotting the data points on special graph paper designed for a particular distributional model. However, methods exist to create probability plots using transformations that can be implemented using spreadsheets or graphics packages on personal computers. Furthermore, many standard statistical packages, such as Minitab, Systat, and others, have built in probability plotting capabilities. If the data are observations obtained from the assumed distribution, then the points should plot as a straight line. If a good fit is obtained, then the parameters of the distribution can be estimated from the probability plot.

Probability paper is most commonly available for the normal and lognormal distributions. If one uses pre-prepared probability paper, the procedure is to: (1) obtain the appropriate type of graph paper; (2) rank the data from smallest to largest observations; (3) estimate the fractiles of the data; (4) plot the value of each data point versus its fractile on the probability paper; (5) inspect the result; and (6) if appropriate, perform a regression analysis.

Probability plots may be generated on the computer using many standard packages, as previously noted. It is also possible to set up probability plots using commonly available spreadsheets. Probability plots require transformations of the data set. In general, a probability plot is based upon a plot of a value Z versus associated values of the random variable, X. The values of Z, sometimes referred to as "Z scores," are calculated based upon the following transformation:

$$z_i = G^{-1}[F_n(x_i)] \tag{5.34}$$

where $G^{-1}(.)$ is the inverse CDF of the standardized distribution for the selected probability distribution model being tested. For a distribution that depends on a location parameter (such as the mean for a normal distribution) and a scale parameter (such as the standard deviation of a normal distribution), the standardized form of the distribution may be written as:

$$z = \frac{x - \mu}{\sigma} \tag{5.35}$$

Thus, as an example, if we have a data set such as the leafy produce PCB concentration data of Chap. 9, and if we use the Hazen method of probability plotting to estimate $F_n(x_i)$, then we can use the inverse CDF for the standardized normal distribution to estimate how many standard deviations away from the mean a given data point should be if in fact it is properly described by a normal distribution. The steps involved in these calculations are summarized in Table 5.6.

The inverse CDF for the normal distribution can be calculated using built-in functions in software such as Excel, or using the following transformation suggested by D'Agostino (1986):

$$t_i = \{-\ln[4F_n(x_i)(1 - F_n(x_i))]\}^{0.5} \tag{5.36}$$

TABLE 5.6 Example of Calculations for a Normal Probability Plot
of the Leafy Produce PCB Concentration Data Set.

Data Set (PCB Conc., ng/g wet basis)	Rank	Plotting Position (rank −0.5)/n	Z-Score from Inverse CDF
0.05	1	0.056	−1.60
0.09	2	0.167	−0.97
0.18	3	0.278	−0.59
0.21	4	0.389	−0.28
0.25	5	0.500	0.00
0.28	6	0.611	0.28
0.28	7	0.722	0.59
0.30	8	0.833	0.97
0.35	9	0.944	1.60

$$z_i = \text{sign}[F_n(x_i) - 0.5][1.238t(1 + 0.0262t)] \qquad (5.37)$$

where the function sign(.) returns +1 if the argument is positive and −1 if the argument is negative. This transformation was used to calculated the values of the Z-score given in Table 5.6. A similar calculation can be done to fit a log-normal distribution to the data by first taking a natural logarithmic transformation of the data, and then proceeding with the rank ordering, plotting position, and Z-score calculations as given in Table 5.6.

If a straight line appears to fit the data on a probability plot, then the chosen probability model may be judged to be acceptable. You should expect that there will be some deviation of data points from a straight line due to random sampling fluctuations, even if the distributional assumption is correct. Furthermore, the deviation will tend to be greater for the tails than for the central values of the distribution. However, if there are systematic deviations from a straight line, then the assumed model is probably not appropriate. How much of a deviation constitutes a rejection of the proposed model is a purely subjective matter and depends in part on the sample size and the data quality objectives for your analysis.

5.7.3. Identifying Distribution Types Based upon Probability Plotting

To illustrate how different data sets will appear on normal and lognormal probability plots, data sets of sample size 50 and 500 have been synthetically generated using the Crystal Ball program. Six distributions were considered: uniform, normal, lognormal, gamma, Weibull, and exponential. Except for the Weibull, all of the distributions were simulated with the same mean and standard deviation. In the case of the Weibull, the relationship between the parameters and the mean and variance is more complex and difficult to quantify. Therefore, an arbitrary shape parameter of two and scale parameter of one were used. The results for the plots of each of the six distributions on normal probability plots are shown in Figs. 5.24 and 5.25 for the cases of $n = 50$ and $n = 500$ data

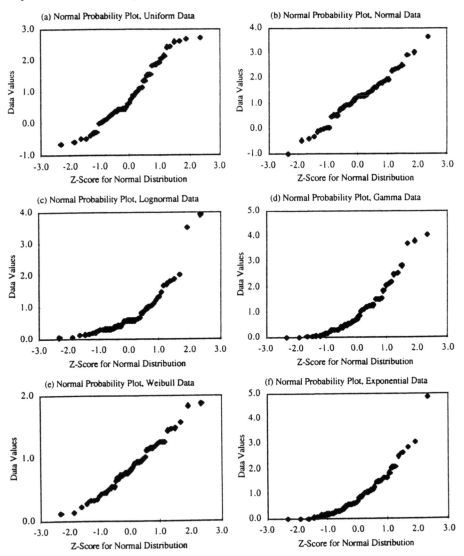

FIGURE 5.24 Normal probability plots for examples of six types of distributions and $n = 50$: uniform, normal, lognormal, gamma, Weibull, and exponential.

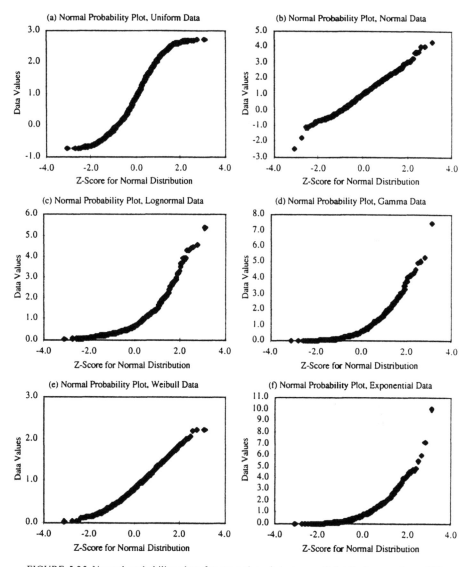

FIGURE 5.25 Normal probability plots for examples of six types of distributions and $n = 500$: uniform, normal, lognormal, gamma, Weibull, and exponential.

points, respectively. The results for lognormal probability plots are shown in Figs. 5.26 and 5.27 for the cases of $n = 50$ and $n = 500$ data points, respectively.

The results indicate some of the characteristic shapes that you may observe depending on the underlying distribution for the data set. On a normal probability plot, a uniformly distributed data set appears as a symmetric S-shaped curve.

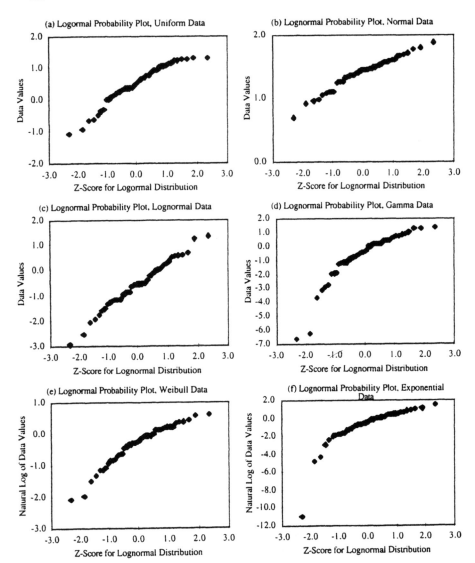

FIGURE 5.26 Lognormal probability plots for examples of six types of distributions and $n = 50$: uniform, normal, lognormal, gamma, Weibull, and exponential.

A normal distribution appears as a straight line. Random sampling error is more apparent for the case of 50 data points than for the case of 500 data points. However, even with 500 data points there can be noticeable fluctuation at the extreme tails. The lognormal, gamma and exponential distributions have similar curves on a normal probability plot. The Weibull distribution appears to plot as a curve

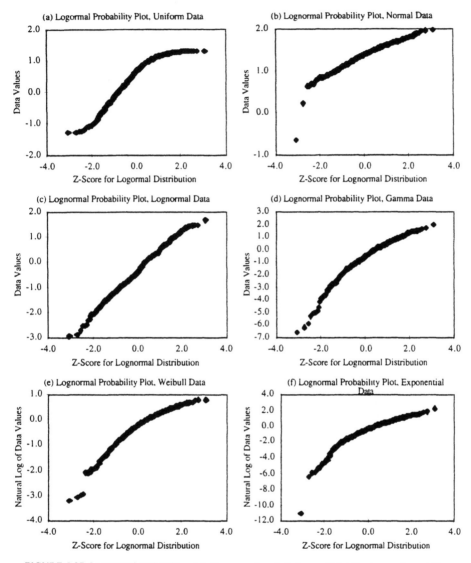

FIGURE 5.27 Lognormal probability plots for examples of six types of distributions and $n = 500$: uniform, normal, lognormal, gamma, Weibull, and exponential.

for small values and as a nearly straight line for larger values, implying that the upper tail may be similar to the normal distribution in this particular case.

The same data are displayed on lognormal probability plots in Figs. 5.26 and 5.27. With only 50 data points, it is difficult to discern a definitive pattern for the uniformly distributed data. However, with 500 data points, it is clearer that the Uniform distribution appears as an asymmetric S-shaped curve. The normal distribution appears to differ from the lognormal most clearly at the lower tail for the case of 500 data points, but appears to be a nearly straight line for the case of 50 data points. This result demonstrates that sampling error due to finite sample sizes may sometimes lead to potentially misleading insights based upon probability plots. As expected, the lognormal data plot as a nearly straight line. The gamma, Weibull, and exponential distributions have similar curves on the lognormal probability plots.

References such as Hahn and Shapiro (1967) and D'Agostino (1986) provide additional examples of data sets plotted on probability paper to illustrate how random sampling can lead to fluctuations and how data which are obtained from other distributions, will appear when plotted on normal probability paper. It is important to keep in mind that, because data for probability plots are rank ordered, they are not independent and therefore should not be expected to be randomly scattered above or below the best fit line. Even for a correct model, there may be a series of successive points above or below the line.

Mixture distributions (briefly mentioned in Chap. 4) may be encountered in analyzing data. A mixture distribution can be decomposed into two or more probability distributions. If a data set appears to be multimodal, then it is likely that a mixture distribution would be an appropriate way to model the data. An example of a probability plot of a mixture distribution is shown in Fig. 5.28. In this case, the data were generated from two different distributions. Of the 100 data points shown, 50 were simulated from a normal distribution with a mean of one and a standard deviation of one. The other 50 were obtained from a normal distribution with a mean of five and a standard deviation of two. As would be expected in this case, the probability plot indicates that the data can be repre-

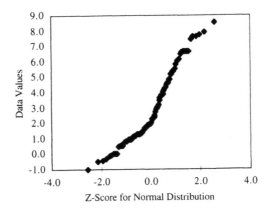

FIGURE 5.28 Normal probability plot for a mixture of two normal distributions: $N(1,1)$ and $N(5,.2)$.

sented by two separate straight lines and that for either component of the mixture, a normal distribution appears to be an adequate model.

The inferences from graphical techniques will fall into the following categories: (1) the assumed probability distribution model appears to be adequate; (2) the model is questionable; and (3) the model is clearly inadequate. Note that an assumed model cannot be *proven* to be adequate.

5.7.4. Normal and Lognormal Probability Plots

For a normal distribution, estimates of the distribution's mean and variance are obtained from the probability plot. Because a normal distribution is symmetric, the mean and median are the same; therefore, the estimate of the 50th percentile serves as an estimate of the mean. The standard deviation can be estimated as 0.4 times the range from the 10th to the 90th percentiles of the distribution. Alternatively, linear regression analysis can be used to fit a straight line to the data. The intercept is the mean, and the slope is the standard deviation. The coefficient of determination provides an indication of the goodness of the fit. It should be noted that least squares linear regression is not appropriate for use with probability plots. This is because rank-ordered data are no longer independent, as previously discussed. Generalized least squares regression should, in principle, be used. However, Hahn and Shapiro (1967) point out such additional effort is often not justified when the final result depends more heavily on subjective judgments regarding selection of appropriate models than upon the precision of the regression method. We will also show by example that least squares linear regression yields good results when the fit is very good.

Examples of normal and lognormal probability plots, based on data sets given in Chap. 9, are shown in Figs. 9.2, 9.3, and 9.4 for PCB concentrations in root, vine, and leafy produce. Each figure displays both normal and lognormal probability plots. In most cases, the graphs indicate that a lognormal distribution typically provides a better fit to the data than does a normal distribution. In some cases, neither distribution appears to be a good fit, such as for leafy produce data. These data sets range from nine to 25 data points. Thus, one expects there may be random sampling error that would make even a good fit look potentially poor.

5.7.5. Estimating Distribution Parameters Using Probability Plots

The parameter estimates obtained from probability plots will often differ from those obtained from other methods. For example, we consider the examples of probability plots given in Chap. 9 and discussed in the previous section. From the probability plots, the parameters of the distributions and the coefficient of determination of the best fit line were calculated. These results are summarized in Table 5.7. Both the method of moments and the probability plot methods yield the same results for the mean for both normal and lognormal probability plots. However, the estimate of the standard deviation differs depending upon the method used. The probability plot method yields estimates of standard deviation less than or equal to that of the method of moments. The methods tend to agree when the coefficient of determination (R^2) is very high (e.g., above 0.9 or 0.95).

When the hypothesized distribution is a poor fit to the data, probability plotting can yield misleadingly low estimates of the standard deviation. For example, in the case of a normal distribution fitted to the outdoor air fall/spring concentration data, the standard deviation based upon the best fit line is 16% less than that calculated directly from the data set. For the seven data sets considered here, all but five appear to be better described by a lognormal than a normal distribution. The exceptions are the leafy produce and winter outdoor air data. In the former case, neither probability plot yields definitive insight regarding selection of the appropriate distribution. In the latter case, there appears to be significant random error in the data. Thus, the selection of a parametric distribution in these cases is more clearly a matter of judgment. There are reasons to suspect that a lognormal distribution may be better in both cases. This is especially so for the leafy data since the coefficient of variation is large enough to lead to a non-negligible probability of obtaining negative values if a normal distribution is assumed.

TABLE 5.7 Comparison of Estimated Mean and Standard Deviation for Selected Data Sets and Log-Transformed Data Sets: Method of Moments and Probability Plots

	Method of Moments		Probability Plot		
Data Set	Mean	Standard Deviation	Mean	Standard Deviation	R^2
Normal Distribution					
Vine Produce (ng PCB/g, wet)	0.10	0.05	0.10	0.04	0.82
Root Produce (ng PCB/g, wet)	0.19	0.13	0.19	0.12	0.85
Leafy Produce (ng PCB/g, wet)	0.22	0.10	0.22	0.10	0.94
Outdoor Air, Summer (ng/m³)	18.84	9.23	18.84	8.90	0.91
Outdoor Air, Winter (ng/m³)	0.83	0.24	0.83	0.24	0.96
Outdoor Air, Fall/Spring (ng/m³)	10.68	15.31	10.68	12.81	0.69
Indoor Air (ng/m³)	20.01	13.66	20.01	12.13	0.78
Lognormal Distribution					
Vine Produce ln(ng PCB/g, wet)	−2.40	0.44	−2.40	0.43	0.96
Root Produce ln(ng PCB/g, wet)	−1.84	0.64	−1.84	0.64	0.98
Leafy Produce ln(ng PCB/g, wet)	−1.65	0.64	−1.65	0.59	0.82
Outdoor Air, Summer ln(ng/m³)	2.83	0.48	2.83	0.48	0.98
Outdoor Air, Winter ln(ng/m³)	−0.23	0.32	−0.23	0.31	0.92
Outdoor Air, Fall/Spring ln(ng/m³)	1.39	1.55	1.39	1.51	0.94
Indoor Air ln(ng/m³)	2.82	0.59	2.82	0.57	0.93

5.7.6. Probability Plots for the Weibull Distribution

In the examples just described, several methods for estimating distribution parameters could be easily compared. For some types of distributions, probability plotting is the most convenient method to use to estimate the distribution parameters. This is true of the Weibull, for which the relationship between the parameters and the first central moments requires evaluation of the gamma function and, hence, can be inconvenient. However, it is relatively easy to develop probability paper for the Weibull. The PDF for the Weibull is (equivalent to Eq. 4.10):

$$f(x) = \frac{k}{c}\left(\frac{x}{c}\right)^{k-1} \exp\left[-\left(\frac{x}{c}\right)^{k}\right] \tag{5.38}$$

where k is a shape parameter and c is a scale parameter. If the Weibull is suspected as being a potentially good fit to a data set, then the following transformation may be used to plot the data:

$$\ln\left\{\ln\left[\frac{1}{\bar{F}(x_i)}\right]\right\} = k \ln(x_i) - k \ln(c) \tag{5.39}$$

where:

$$\bar{F}(x_i) = 1 - F(x_i) = Pr(X > x_i) \tag{5.40}$$

is the complementary CDF of X. Thus, using most spreadsheet or graphics packages, it is possible to plot the data set and to evaluate the scale and shape parameters using conventional least-squares regression techniques. An example is shown in Fig. 5.29 for $n = 100$. In this example, the best fit equation was found to be:

$$\ln\left\{\ln\left[\frac{1}{\bar{F}(x_i)}\right]\right\} = 1.8084 \ln(x_i) + 0.0736 \tag{5.41}$$

The shape parameter can be evaluated by inspection as $k = 1.808$. The scale parameter can be found by solving the following expression:

$$c = \exp\left(\frac{-0.0736}{k}\right) \tag{5.42}$$

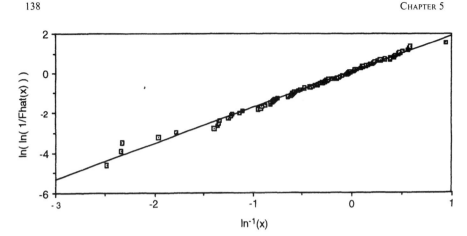

FIGURE 5.29 Example of a probability plot for a Weibull distribution.

to obtain $c = 0.96$. As an aside, the data used for this example were generated from a parent distribution with a shape parameter of two and a scale parameter of one. Thus, there can be significant sampling error in the estimates of these parameters.

5.7.7. Probability Plots and Censored Data Sets

Graphical techniques are particularly useful for dealing with censored distributions. A censored distribution is one for which data are not available above or below a certain cut-off value. For example, concentration data may not be available below a specified "detection-limit." In such cases, you can plot the available data on probability paper and evaluate the reasonableness of a distributional assumption. If the data agree well with the assumption, it may then be plausible to extrapolate the distribution into the censored region.

A hypothetical example is given here to illustrate the approach. Suppose we have 500 data points from a parent Weibull distribution. A fitted distribution may be developed as shown in Fig. 5.30. Suppose 10% of the data points at the lower end of the data set were actually nondetects. Then the remaining 450 data points would be used to estimate the parameters of the Weibull. However, in calculating the complementary CDF (CCDF) in Eq. (5.40), the information contained in the nondetects would be used. Specifically, no values of the CCDF would be plotted for fractiles of less than 0.0998 ($50/(n + 1)$, where $n = 500$). The resulting data and best fit line are shown in Fig. 5.31. The estimates of the parameters of the distribution differ modestly for the case with censored data versus the case with all data. In the former case, the estimates of the parameters are 1.874 for the shape and 0.969 for the scale. In the latter case, the estimates are 1.777 and

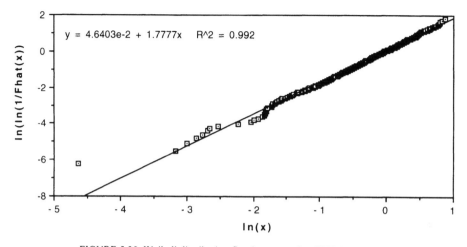

FIGURE 5.30 Weibull distribution fitted to a sample of 500 data points.

0.974, respectively. Thus, the estimates of parameters for censored distributions may differ from those that would be obtained if all data had been available.

5.7.8. Percentile-Percentile Plots

The probability plotting methods illustrated to this point are examples of quartile–quartile (Q–Q) plots. There are other graphical methods for evaluating

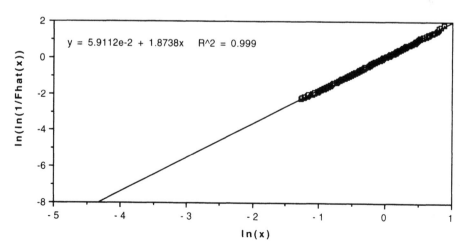

FIGURE 5.31 Weibull distribution fitted to a censored sample of 450 data points.

the fit of a distribution to data. Some of these include the percentile–percentile (P–P) plot and frequency comparisons.

A Q–Q plot involves plotting the quantiles (rank-ordered data values) of one distribution versus the quantiles of another distribution. In the case of the probability plots presented in previous examples, the quantiles used for the other distribution have been based upon a standardized random variable for the hypothesized distribution against which the data are being evaluated. Thus, a probability plot is a special case of a Q–Q plot (D'Agostino, 1986). A Q–Q plot can be constructed without having to estimate the parameters of the distribution for the types of distributions that are described by a location and scale parameter. Furthermore, the insight regarding the best fit using a Q–Q plot will be unaffected by linear transformations of the data, although such transformations will affect the estimate of the parameters based upon a best fit line. Thus, Q–Q plots are said to be location and scale equivariant (Gan and Koehler, 1990). Q–Q plots tend to emphasize discrepancies in the fit of the tails of the distribution to the data set (Gan *et al.*, 1991).

An example of a P–P plot is shown in Fig. 5.32. In this figure, the empirically observed percentiles of each depicted data set are given on the horizontal

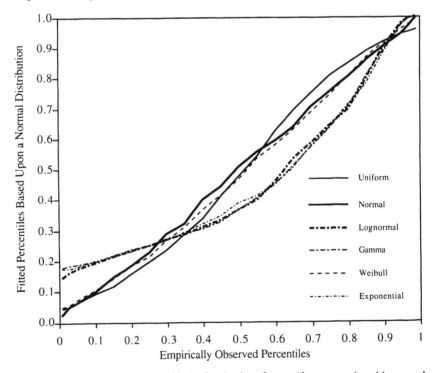

FIGURE 5.32 Normal P–P plot of 500 simulated values from uniform, normal, and lognormal, gamma, Weibull, and exponential distributions.

axis. For each data set, the mean and standard deviation were calculated. These values were in turn used to estimate the percentile of a standard normal distribution that would be associated with each of the observed data points if the data were described by a normal distribution. If the data are normally distributed, then they should fall on a 45 degree line. Thus, in contrast to the Q–Q plot, for which a reference line must be developed for each individual case, the goodness-of-fit on a P–P plot can be evaluated with respect to any deviation from a standard 45 degree reference line. The trade-off is that one must estimate parameters in order to develop the P–P plot, whereas it is not necessary to do so in order to develop a Q–Q plot. In contrast to the Q–Q plot, the standardized P–P plot is invariant with respect to linear transformations of the data. Furthermore, whereas the Q–Q plot is sensitive to discrepancies at the tail, the P–P plot is more sensitive to discrepancies near the median (D'Agostino, 1986; Gan and Koehler, 1990; Gan et al., 1991). Thus, a Q–Q plot may be more useful for evaluating the fit of a distribution to data when one is mostly concerned with the tails. The P–P plot may be more useful when one is concerned with the central tendency of the distribution.

Like the Q–Q plots, data that result from different underlying distributions appear as characteristic shapes on a standardized normal P–P plot. In Fig. 5.32, as expected, the normally distributed data appear as approximately a straight line with a slope of one. All symmetric distributions should pass through the (0.5,0.5) point on the graph. For example, the uniform distribution passes approximately through this point. However, since the uniform distribution differs in kurtosis from the normal, it does not appear as a straight line. Instead, it has a characteristic S-shaped curve. Skewed distributions do not pass through the (0.5,0.5) point. Positively skewed distributions will intersect the 45 degree reference line at some point below the median value of the distribution, and negatively skewed distribution will intersect the 45 degree line above the (0.5,0.5) point. The lognormal, gamma, and exponential distributions all display the characteristics of positively skewed distributions. In this particular example, the Weibull distribution is approximately symmetric, and therefore appears to be similar to a normal distribution. Gan and Koehler (1990) discuss in more detail the characteristic shapes one would expect to see on a P–P plot, depending on the nature of the underlying distribution.

An example of a P–P plot for a data set of 50 samples drawn from a normal distribution is shown in Fig. 5.33. This example illustrates the type of insights provided by the P–P plot. The fit near the tails appears to be very good. However, the discrepancies in the central values of the distribution are more pronounced than in the Q–Q probability plot. This example also illustrates that plots of rank ordered data are no longer statistically independent representations of the data. Although there appears to be scatter above and below the reference line, there are significant "runs" of consecutive values above the line, followed by a run of consecutive values below the line.

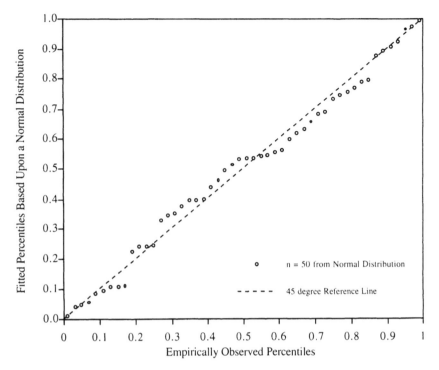

FIGURE 5.33 Standardized normal P–P plot of a normally-distributed data set.

5.7.9. Frequency Comparisons

Another type of graphical method for comparing data and distributions is shown in Fig. 5.34. This is a frequency comparison in which a histogram of a data set is compared with a probability density function. The PDF was calculated based upon parameter estimates from the data set. In this example, 500 data

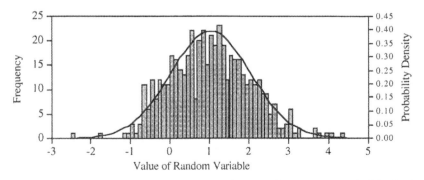

FIGURE 5.34 An example of a frequency comparison between a histogram and a probability density function.

values drawn at random from a normal distribution are depicted. The histogram illustrates that the empirical estimate of the PDF, in the form of the histogram, can be rather "noisy." The smooth curve of the PDF calculated from the mean and standard deviation of the data appears to agree with the histogram, thus indicating a reasonable fit to the data.

5.8. GOODNESS-OF-FIT TESTS

There are many goodness-of-fit tests. Two general types of approaches for evaluating the goodness of fit of a hypothesized distribution with respect to a data set include probability plots and statistical tests. It is widely recognized that probability plots are a subjective method for determining whether or not data contradict an assumed model based upon a visual inspection. However, some statistical methods can be used in conjunction with probability plots to provide a numerical indication of the goodness-of-fit. These include, for example, the use of regression techniques, as previously discussed. In this section, we will focus upon the use of statistical techniques.

While many statistical techniques exist, we will examine only three to provide some examples of the variety of methods. The reader interested in more information may wish to consult a reference such as D'Agostino and Stephens (1986), which contains a far more comprehensive review of statistical techniques for evaluating goodness-of-fit. The three techniques we will focus on include the chi-squared test, the Kolmogorov–Smirnov test, and the Anderson–Darling test.

Note that most statistical tests may only be employed if a minimum amount of data are available. In some cases, such as for the chi-squared test, at least 25 data points should be available. The Kolmogorov–Smirnov test can be used with data sets with as few as five data points. The Anderson–Darling test is valid if the number of samples is greater than or equal to eight. With all of these tests, a test statistic is calculated. The value of the test statistic is then compared to the distribution for the test statistic. If the value obtained from the data analysis falls within the range of likely values of the test statistic, then the null hypothesis that the data were obtained from the hypothesized distribution cannot be rejected. However, if the value of the statistic exceeds reasonable values of the distribution of the test statistic, then the null hypothesis is rejected and, therefore, the hypothesized distributed is rejected. The level at which one distinguishes between likely and unlikely values of the test statistic is a matter of judgment. Typically, a significance level of 0.05 is used, implying that a value of the test statistic below the 95th percentile of the distribution for the statistic is acceptable and leads to an inability to reject the hypothesis. A value of the test statistic above the 95th percentile of the distribution for the test statistic leads to rejection of the hypothesis. However, you can choose other significance levels. The selection of the significance level is one of the subjective aspects of goodness-of-fit tests. Furthermore, note that it is never possible to prove that a hypothesized distribution is "correct."

5.8.1. Chi-Squared Test

The chi-squared test is highly versatile. This test can be used to test any distributional assumption. The chi-squared test involves calculating a test statistic that approximately follows a chi-square distribution only if the hypothesized model cannot be rejected as a poor fit to the data. The test statistic will tend to exceed acceptable values of the chi-square distribution if the hypothesized model is not correct.

If we have a data set for which we know the sample values, the chi-squared test involves: (a) selecting a hypothesized distribution; (b) estimating the parameters of the distribution from the data set; (c) grouping the values into cells (or bins) in which each cell has at least five data points; (d) calculating the probability of obtaining values within the range of each cell based upon the hypothesized distribution; (e) calculating the expected number of data points that should be in each cell if the hypothesized distribution is acceptable; (f) calculating a test statistic; and (g) evaluating the test statistic. We will demonstrate each of these steps using an example.

In the first step, we will use a simulated data set of 500 values drawn from a normal distribution with mean of one and standard deviation of one. This example has been used previously in discussing probability plots. We hypothesize that the data set is drawn from a normal distribution. The estimates of the mean and standard deviation of the data set, which are 1.03 and 1.02, respectively, are used as estimates for the parameter values of the hypothesized normal distribution. The 500 data values are then grouped into cells. The number of cells that should be used is a matter of judgment. A rule of thumb is (Hahn and Shapiro, 1967):

$$
\text{for } n < \sim 200; \ k \leq \frac{n}{5}
$$
$$
\text{for } n > \sim 200; \ k = \text{integer}\{4[0.75(n-1)^2]^{0.2}\}
$$

(5.43)

where k = the number of cells. Thus, in this example, since $n = 500$, we calculated that the number of cells should be approximately 45. However, since 500 divided by 45 does not yield an integer value, we will use 50 cells. With smaller data sets, it is often desirable to have as many cells as possible, subject to the constraint that no cell should have less than five data points.

It is convenient but not necessary to group the data into cells such that there is equal probability of sampling data within any given cell. One can also group data in a way so that each cell has an equal range of values, which typically will correspond to different probabilities of obtaining values in one cell versus another. In the example, we have divided the data into 50 cells with 10 data points each, as shown in Table 5.8. The probability that the values in any given cell would be sampled from the assumed normal distribution was calculated. This was done by calculating the value of the CDF for the normal distribution with mean of 1.03 and standard deviation of 1.02 at the upper bound value of each cell range, and subtracting the value of the CDF for the lower bound value of each cell range. The expected number of data points for each cell was then calculated by multiplying the probability of each cell by the total number of data points. A test statistic is then computed as follows:

$$X^2 = \sum_{i=1}^{k} \frac{(M_i - E_i)^2}{E_i} \qquad (5.44)$$

where M_i is the number of data values in each cell, and E_i is the expected number of data values in each cell.

In the example, the test statistic has a value of 52.1. We compare this with a value of the chi-square distribution with $k - r - 1$ degrees of freedom, where r is the number of parameters that were estimated for the hypothesized distribution. Since the normal distribution has two parameters, $r = 2$. Therefore, in this case, we have 47 degrees of freedom. We can choose any confidence level we wish for the chi-square distribution. Typically, a 95% confidence level (5% significance level) is chosen. The corresponding value from the chi-square distribution is 64.00. Therefore, since the value of our test statistic is less than the 95 percentile value of the chi-square distribution with 47 degrees of freedom, we cannot reject the hypothesized model as being inadequate on the basis of this test.

We will now consider the use of the chi-squared test for one of the data sets of the case study in Chap. 9. We focus here on the root produce data set, for which $n = 25$. These data were grouped into five cells, each with five data points, as indicated in Table 5.9. The probability of obtaining values within each cell was calculated based upon a hypothesized normal distribution, with mean of 0.19 and standard deviation of 0.13. The test statistic was calculated based upon the expected number of values in each cell versus the actual number in each cell. The test statistic has a value of 10.3. This is larger than the 95th percentile of a chi-square distribution with two degrees of freedom, which has a value of 6.0. On this basis, we may reject the hypothesized normal distribution as an adequate fit to the data. We observe from the table that most of the deviation from normality appears to be occur in the first and second cells.

We next consider whether a lognormal distribution may fit the root produce data. In fact, we have more reason to believe that a lognormal distribution is appropriate, since the data are for a quantity which must be non-negative. The

TABLE 5.8 Example of Chi-Squared Test for a Simulated Normally-Distributed Data Set of $n = 500$.

Cell Number	End Points of Each Cell		Number of Values in Cell, M_i	Cell Probability, p_i, Based on Normal Distribution	Expected Number of Values in Cell, E_i	Test Statistic
	Lower Bound	Upper Bound				
1	-2.48	-0.75	10	0.0396	19.80	4.85
2	-0.75	-0.61	10	0.0128	6.39	2.05
3	-0.61	-0.48	10	0.0151	7.57	0.78
4	-0.48	-0.41	10	0.0105	5.24	4.33
5	-0.41	-0.29	10	0.0182	9.12	0.08
6	-0.29	-0.21	10	0.0137	6.85	1.45
7	-0.21	-0.11	10	0.0200	9.99	0.00
8	-0.11	0.00	10	0.0237	11.86	0.29
9	0.00	0.04	10	0.0120	5.99	2.68
10	0.04	0.11	10	0.0162	8.09	0.45
11	0.11	0.19	10	0.0222	11.12	0.11
12	0.19	0.23	10	0.0112	5.60	3.46
13	0.23	0.31	10	0.0236	11.81	0.28
14	0.31	0.38	10	0.0225	11.27	0.14
15	0.38	0.45	10	0.0217	10.84	0.07
16	0.45	0.51	10	0.0216	10.79	0.06
17	0.51	0.54	10	0.0093	4.64	6.18
18	0.54	0.60	10	0.0224	11.21	0.13
19	0.60	0.71	10	0.0409	20.47	5.35
20	0.71	0.77	10	0.0215	10.74	0.05
21	0.77	0.81	10	0.0147	7.34	0.96
22	0.81	0.86	10	0.0211	10.57	0.03
23	0.86	0.89	10	0.0131	6.53	1.84

24	0.89	0.95	10	0.0205	10.23	0.01
25	0.95	1.04	10	0.0351	17.56	3.25
26	1.04	1.08	10	0.0183	9.13	0.08
27	1.08	1.11	10	0.0120	6.02	2.62
28	1.11	1.17	10	0.0230	11.50	0.20
29	1.17	1.24	10	0.0246	12.30	0.43
30	1.24	1.26	10	0.0110	5.48	3.72
31	1.26	1.31	10	0.0175	8.74	0.18
32	1.31	1.37	10	0.0223	11.15	0.12
33	1.37	1.41	10	0.0163	8.17	0.41
34	1.41	1.50	10	0.0303	15.17	1.76
35	1.50	1.56	10	0.0229	11.46	0.19
36	1.56	1.63	10	0.0223	11.13	0.11
37	1.63	1.68	10	0.0156	7.81	0.61
38	1.68	1.76	10	0.0246	12.30	0.43
39	1.76	1.81	10	0.0147	7.34	0.97
40	1.81	1.89	10	0.0227	11.34	0.16
41	1.89	1.97	10	0.0214	10.71	0.05
42	1.97	2.06	10	0.0220	10.99	0.09
43	2.06	2.15	10	0.0208	10.38	0.01
44	2.15	2.24	10	0.0187	9.36	0.04
45	2.24	2.34	10	0.0176	8.79	0.17
46	2.34	2.43	10	0.0153	7.63	0.74
47	2.43	2.59	10	0.0216	10.81	0.06
48	2.59	2.77	10	0.0189	9.46	0.03
49	2.77	3.08	10	0.0209	10.43	0.02
50	3.08	4.32	10	0.0207	10.35	0.01
Sum of Values:			500	0.999	499.58	52.10

TABLE 5.9 Example of Chi-Squared Test of a Hypothesized Normal Distribution (Mean = 0.19, Standard Deviation = 0.13) for Root Produce PCB Concentration Data Set of n = 25.

Cell Number	End Points of Each Cell		Number of Values in Cell, M_i	Cell Probability, p_i, Based on Normal Distribution	Expected Number of Values in Cell, E_i	Test Statistic
	Lower Bound	Upper Bound				
1	0.04	0.09	5	0.0941	2.35	2.98
2	0.09	0.12	5	0.0732	1.83	5.50
3	0.12	0.17	5	0.1432	3.58	0.56
4	0.17	0.27	5	0.2955	7.39	0.77
5	0.27	0.51	5	0.2699	6.75	0.45
Sum of Values:			25	0.876	21.90	10.27

results of the chi-squared test for a hypothesized lognormal distribution are shown in Table 5.10. The approach employed here was to take the natural logarithm of each data point in the root produce PCB concentration data set, and then to evaluate the fit of a normal distribution to the logarithmic-transformed data. This is equivalent to evaluating the fit of a lognormal distribution to the original data. The test statistic was found to be 1.12, which is well below the 95th percentile value of 6.0 based upon two degrees of freedom. Thus, we cannot reject, on the basis of this test, the lognormal distribution as an adequate fit to the data set.

While we have focused on the application of the chi-squared test to cases involving normal or lognormal distributions, it can be used for other distributions as well. Thus, an advantage of this method is its flexibility. A disadvantage is that it has a lower power than many other statistical tests. This is because the chi-squared test requires grouping of data. As a result of the groupings, some information regarding the original data set is lost.

5.8.2. Kolmogorov–Smirnov Test

The Kolmogorov–Smirnov test involves a comparison between a stepwise empirical CDF and the CDF of the hypothesized distribution. The maximum discrepancy in the estimated cumulative probabilities for the two CDFs is identified. The maximum discrepancy is then compared to a critical value of the test statistic. If the maximum discrepancy is larger than the critical value, then the hypothesized distribution is rejected. This method is illustrated with several examples involving data sets from the case study of Chap. 9. This method is discussed by Ang and Tang (1975) and others.

First, we consider the leafy produce PCB concentration data set. In previous discussions of the central moments and probability plots for this data set, it has been difficult to make a conclusion regarding the underlying probability distribution. We did not apply the chi-squared test to this data set, because we have only nine data points. However, an advantage of the Kolmogorov–Smirnov test is that it can be done with much fewer data points than can the chi-squared test. We begin by constructing the required stepwise cumulative probability function, $S_n(x)$, where n = the number of data points. First, we rank order the n data points so that $x_1 < x_2 < \cdots < x_n$. Then, $S_n(x)$ is defined as follows:

$$
S_n(x) = \begin{cases} 0 & \text{for } x < x_1 \\ \dfrac{i}{n} & \text{for } x_i \leq x < x_{i+1} \\ 1 & \text{for } x \geq x_n \end{cases} \tag{5.45}
$$

TABLE 5.10 Example of Chi-Squared Test of a Hypothesized Lognormal Distribution (Mean = −1.84, Standard Deviation = 0.64) for Logarithmically-Transformed Root Produce PCB Concentration Data Set of $n = 25$.

Cell Number	End Points of Each Cell		Number of Values in Cell, M_i	Cell Probability, p_i, Based on Normal Distribution	Expected Number of Values in Cell, E_i	Test Statistic
	Lower Bound	Upper Bound				
1	−3.22	−2.41	5	0.1715	4.29	0.12
2	−2.41	−2.12	5	0.1427	3.57	0.58
3	−2.12	−1.77	5	0.2108	5.27	0.01
4	−1.77	−1.31	5	0.2539	6.35	0.29
5	−1.31	−0.67	5	0.1704	4.26	0.13
Sum of Values:			25	0.949	23.74	1.12

Similarly, the CDF, $F(x)$, for the hypothesized distribution is calculated. This is typically done by selecting a parametric distribution, estimating the parameters of the distribution from the data set, and then determining $F(x)$ based upon the selected distribution and parameter estimates. In Fig. 5.35, $S_9(x)$ for the leafy produce PCB concentration data set is compared to $F(x)$ for a hypothesized normal distribution, whereas in Fig. 5.36, $S_9(x)$ is compared to a lognormal distribution.

The next step in the method is to compute the maximum difference, D_n, between $F(x)$ and $S_n(x)$:

$$D_n = \max_x |F(x) - S_n(x)| \qquad (5.46)$$

D_n is a random variable whose distribution depends upon the sample size n. Values of this test statistic can be looked up in tables in texts such as Ang and Tang (1975). For example, values of D_n at a significance level of 0.05 are 0.56 for $n = 5$, 0.41 for $n = 10$, 0.34 for $n = 15$, 0.29 for $n = 20$, and 0.27 for $n = 25$. For $n > 50$, the value of D_n at a significance level of 0.05 can be estimated as $1.36/\sqrt{n}$.

For the leafy produce data set, the maximum difference between $F(x)$ for a normal distribution and $S_9(x)$ for the empirical stepwise cumulative distribution is 0.17. This value is less than the critical value of the test statistic. At $n = 10$,

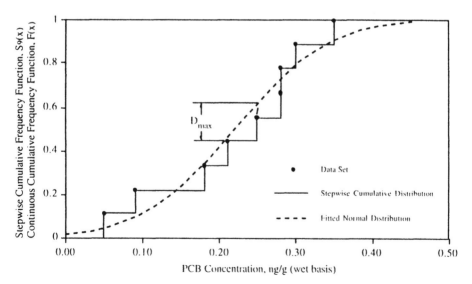

FIGURE 5.35 Empirical stepwise cumulative frequency distribution versus hypothetical normal distribution for leafy produce PCB concentration data set.

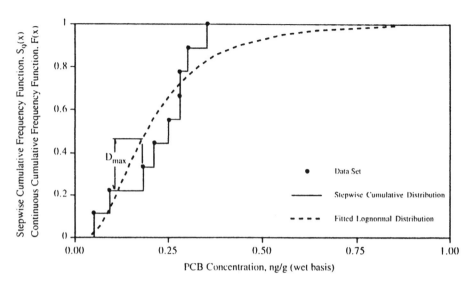

FIGURE 5.36 Empirical stepwise cumulative frequency distribution versus hypothetical lognormal distribution for leafy produce PCB concentration data set.

the critical value of the test statistic is 0.41. Since we have $n = 9$, we would expect the critical value to be somewhat larger than this. Since our calculated value is less than the critical value for $n = 10$, it will also be less than the critical value for $n = 9$. Therefore, we cannot reject the normal distribution as an adequate fit to this data set at the 0.05 significance level. Similarly, for the hypothesized lognormal distribution, $D_9 = 0.24$. This is also less than the critical value for the test statistic. Therefore, in this case, it is not possible to reject either a normal or a lognormal distribution based upon the Kolmogorov–Smirnov test.

Similar analyses were done for the vine produce PCB data set, as shown in Figs. 5.37 and 5.38. For the fitted normal distribution, D_{17} was 0.24, while for the fitted lognormal distribution it was 0.16. For $n = 17$, the critical value of D_{17} at the 0.05 significance level is approximately 0.32. Thus, neither the normal nor lognormal distributions can be rejected on the basis of this test.

For the root produce data set, as shown in Figs. 5.39 and 5.40, the maximum deviation of the stepwise cumulative distribution from the fitted distribution is 0.17 for the normal distribution and 0.08 for the lognormal distribution. For $n = 25$, the critical value of D_{25} is 0.27 at the 0.05 significance level. Thus, it is not possible to reject either the normal or lognormal distribution as an incorrect fit based upon this test. However, it appears to be the case that the lognormal distribution provides a better fit, since the maximum deviation is less than for the normal distribution. The three examples given here illustrate that goodness-of-fit tests often cannot be used to identify a unique parametric distribution that describes the data.

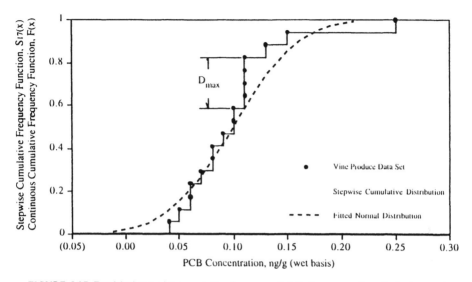

FIGURE 5.37 Empirical stepwise cumulative frequency distribution versus hypothetical normal distribution for vine produce PCB concentration data set.

The Komolgorov–Smirnov test can be modified to deal with cases in which there is censoring of the data set (Stephens, 1986). There are different types of censoring. For example, if all observations less than a value x_s are missing, then the sample is said to be "left-censored." If all observations greater than x_r are

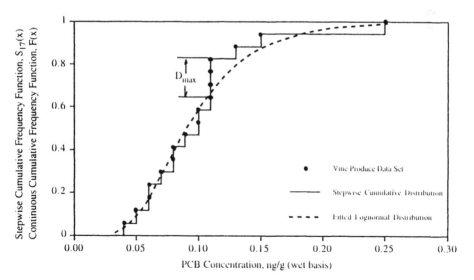

FIGURE 5.38 Empirical stepwise cumulative frequency distribution versus hypothetical lognormal distribution for vine produce PCB concentration data set.

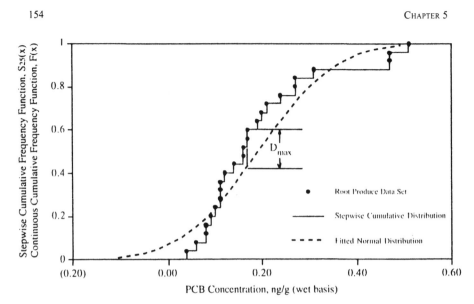

FIGURE 5.39 Empirical stepwise cumulative frequency distribution versus hypothetical normal distribution for root produce PCB concentration data set.

missing, then the sample is said to be "right-censored." If observations are missing at only the low end or only the high end, the sample is said to be "singly censored." If observations are missing at both the high and low ends, then the sample is "doubly censored." The lower and upper limits, x_s and x_r, beyond which

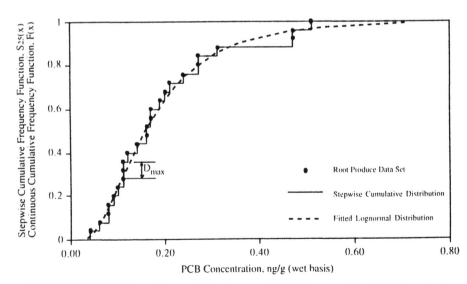

FIGURE 5.40 Empirical stepwise cumulative frequency distribution versus hypothetical lognormal distribution for root produce PCB concentration data set.

censoring occurs may themselves be uncertain quantities, which is known as Type 1 censoring. Alternatively, if x_s and x_r are known fixed quantities, then we have Type 2 censoring. The Kolmogorov–Smirnov test is adjusted for censoring primarily on the basis of whether Type 1 or Type 2 censoring applies. The specifics of the modified tests are given by Stephens (1986).

5.8.3. Anderson–Darling Test

Like the Kolmogorov–Smirnov test, the Anderson–Darling test is an example of an empirical distribution function (EDF) test. The Kolmogorov–Smirnov test is what is known as a "supremum" test, which involves finding the maximum vertical distance between the empirical stepwise CDF and the fitted CDF. In contrast, the Anderson–Darling test is known as a "quadratic" test, because it is based upon a weighted square of the vertical distance between the empirical and fitted CDFs. The Anderson–Darling test uses one type of weighting. Similar tests, such as the Cramer-von Mises statistic and the Watson statistic, employ different weighting schemes (Stephens, 1986).

For evaluating the fit of a normal distribution, the Anderson–Darling test is performed as follows: (a) arrange the data in ascending order; (b) calculate standardized values of the data; (c) calculate the cumulative probability for the fitted distribution; (d) calculate the Anderson–Darling statistic; (e) compute a modified statistic, A^*; and (f) compare the modified statistic to a critical value to decide whether to reject the hypothesis that the data are described by the hypothesized distribution.

We will illustrate the Anderson–Darling test with respect to an application to the root produce PCB concentration data set of Chap. 9, as shown in Table 5.11. The data set was rank ordered, and the standardized values for the normal distribution were calculated as follows:

$$Y_i = \frac{x_i - \bar{x}}{s} \tag{5.47}$$

From the standardized values, the cumulative probability was calculated using the standard normal distribution:

$$p_i = \Phi(Y_i) \tag{5.48}$$

where $\Phi(.)$ is the CDF of the standard normal distribution. The next step is to compute the Anderson–Darling statistic, A^2:

$$A^2 = -\sum_{i=1}^{n} \frac{(2i - 1)[ln(p_i) + ln(1 - p_{n+1-i})]}{n} - n \tag{5.49}$$

TABLE 5.11 Anderson–Darling Test for Goodness-of-Fit of a Normal Distribution
to the Root Produce PCB Concentration Data Set

Rank	Data Value ng/g (wet)	Standardized Value	Cumulative Probability	Test Statistic
1	0.04	−1.19	0.12	−0.28
2	0.06	−1.03	0.15	−0.72
3	0.08	−0.88	0.19	−1.16
4	0.08	−0.88	0.19	−0.94
5	0.09	−0.80	0.21	−1.02
6	0.1	−0.73	0.23	−1.20
7	0.11	−0.65	0.26	−1.24
8	0.11	−0.65	0.26	−1.29
9	0.11	−0.65	0.26	−1.42
10	0.12	−0.57	0.28	−1.47
11	0.14	−0.42	0.34	−1.38
12	0.16	−0.26	0.40	−1.36
13	0.16	−0.26	0.40	−0.43
14	0.17	−0.18	0.43	−1.46
15	0.17	−0.18	0.43	−1.47
16	0.19	−0.03	0.49	−1.30
17	0.2	0.05	0.52	−1.26
18	0.21	0.13	0.55	−1.25
19	0.24	0.36	0.64	−1.10
20	0.27	0.59	0.72	−0.92
21	0.27	0.59	0.72	−0.92
22	0.31	0.90	0.82	−0.71
23	0.47	2.14	0.98	−0.41
24	0.47	2.14	0.98	−0.34
25	0.51	2.45	0.99	−0.26
	Anderson–Darling Statistic			1.31
	Modified Statistic			1.36

The Anderson–Darling statistic is then modified based on the sample size for comparison with the critical value:

$$A^* = A^2 \left(1.0 + \frac{0.75}{n} + \frac{2.25}{n^2} \right) \qquad (5.50)$$

The modified value, A^*, is then compared with a critical value. The critical value depends upon the desired significance level. The values of A^* are 0.631, 0.752, 0.873, 1.035, and 1.159 for the significance levels of 0.10, 0.05, 0.025, 0.01, and 0.005, respectively. This approach is valid for sample sizes greater than or equal to eight (D'Agostino, 1986).

The test for normality as shown in Table 5.11 indicates that the modified value $A*$ is 1.36. This value is substantially higher than the critical values for any of the significance levels considered. Therefore, we reject the hypothesis that a normal distribution adequately describes this data set. We also consider a test for lognormality, based upon a logarithmic transformation of the data. This is given in Table 5.12. The resulting modified test statistic is 0.182, which is well within the critical values of the test statistic for all significance levels considered.

The Anderson–Darling test was also applied to the vine and leafy produce PCB concentration data sets to evaluate both fitted normal and lognormal distributions. For the vine data set, the modified statistics for the fitted normal and lognormal distributions were 0.872 and 0.309, respectively. These values indi-

TABLE 5.12 Anderson–Darling Test for Goodness-of-Fit of a Lognormal Distribution to the Root Produce PCB Concentration Data Set

Rank	Data Value ng/g (wet)	Standardized Value	Cumulative Probability	Test Statistic
1	–3.22	–2.15	0.02	–0.30
2	–2.81	–1.52	0.06	–0.70
3	–2.53	–1.07	0.14	–1.01
4	–2.53	–1.07	0.14	–1.08
5	–2.41	–0.89	0.19	–1.17
6	–2.30	–0.72	0.23	–1.33
7	–2.21	–0.58	0.28	–1.36
8	–2.21	–0.58	0.28	–1.42
9	–2.21	–0.58	0.28	–1.55
10	–2.12	–0.44	0.33	–1.56
11	–1.97	–0.20	0.42	–1.38
12	–1.83	0.01	0.50	–1.35
13	–1.83	0.01	0.50	–1.39
14	–1.77	0.10	0.54	–1.42
15	–1.77	0.10	0.54	–1.35
16	–1.66	0.28	0.61	–1.11
17	–1.61	0.36	0.64	–1.03
18	–1.56	0.43	0.67	–1.03
19	–1.43	0.64	0.74	–0.94
20	–1.31	0.82	0.79	–0.78
21	–1.31	0.82	0.79	–0.72
22	–1.17	1.04	0.85	–0.54
23	–0.76	1.69	0.95	–0.36
24	–0.76	1.69	0.95	–0.21
25	–0.67	1.81	0.97	–0.10
	Anderson–Darling Statistic			0.176
	Modified Statistic			0.182

cate that the normal distribution should be rejected at the 0.05 significance level, although it can be accepted at the 0.025 significance level, while the lognormal distribution can be accepted for all of the significance levels considered. For the leafy produce data set, the modified test statistic was 0.361 for the normal distribution and 0.815 for the lognormal distribution. These values indicate that the null hypothesis of normality cannot be rejected for the leafy data set. However, they also indicate that the null hypothesis of lognormality cannot be rejected for significance levels of 0.025 or lower. Thus, the selection of an appropriate distribution remains a matter of judgment in this case.

In the sections on probability plotting and goodness-of-fit testing, we have highlighted the ambiguities and subjective aspects of applying these techniques. Both approaches involve subjective judgment regarding what constitutes an acceptable fit. Furthermore, different methods may emphasize different aspects of a distribution. Conventional Q–Q probability plots emphasize deviations at the tails, while P–P plots emphasize deviations in the central tendency. Chi-squared tests are highly versatile, but require judgment in setting up the cells and in the selection of significance levels for evaluation of the test statistic. The Kolmogorov–Smirnov and Anderson–Darling tests also involve subjective judgment regarding the choice of significance levels. As shown in some cases, a given test may indicate that two or more distributions provide acceptable fits to a distribution. Thus, probability plots and statistical tests may be an insufficient basis for making decisions regarding which distributions to fit to a given data set.

Many authors emphasize the subjective nature of statistical tests. Hahn and Shapiro (1967) state this quite well in their excellent book:

> One might conclude . . . that a proper procedure for selecting a distribution is to consider a wide variety of possible models, evaluate each by the methods here described, and assume as correct the one that provides the best fit to the data. However, *no* such approach is being suggested. Where possible, the selection of the model should be based on an understanding of the underlying physical properties. . . . The distributional test then provides a useful mechanism for evaluating the adequacy of the physical interpretation. Only as a last resort is the reverse procedure warranted, and then, only with much care, for, although many models might appear appropriate within the range of the data, they might well be in error in the range for which predictions are desired. [pp. 260–261]

It is an often repeated fallacy that statistical tests are "objective." They are not. At best, they can be said to be empirically based. However, often more important than the apparent patterns in a data set are the underlying processes that generated the data. In many cases, knowledge of the latter is more useful in selecting distribution models than is inspection of data. For example, for the leafy produce PCB concentration data sets, nearly all of the many analyses we have presented here point to the normal distribution as being a better fit to the data than the lognormal distribution. However, the normal distribution is fundamentally inappropriate in this case. This is because PCB concentrations cannot be negative, and yet the use of a normal distribution would lead to this type of prediction. Furthermore, by analogy with similar data sets, such as for vine and root produce, we may infer that a lognormal distribution is reasonable. Thus, for these reasons, a lognormal distribution has been assumed for this data set in the case study of Chap. 9.

CHAPTER 6

SPECIAL TOPICS RELATED TO DISTRIBUTION DEVELOPMENT

In this chapter we consider special topics related to distribution development. We discuss initial or default distributions, maximum entropy inference as an alternative to traditional statistical approaches, the complexities of combining information, combating surprise, accounting for correlation between inputs, and very briefly, expert elicitation.

6.1. INITIAL DISTRIBUTIONS

The surge in popularity of probabilistic methods for exposure assessment has prompted the publication of a host of initial (or default) distributions for various exposure factors. These distributions appear in documents released by private companies and government agencies (ICRP, 1975; USEPA, 1989; AIHC, 1994; USEPA, 1996a), as well as in many sources in the primary literature[1]. The distributions found in the published papers of individual researchers usually are generated for specific, carefully-defined purposes and in general are not suggested for use outside the context stated by the authors. However, any available distribution may be intriguing to an analyst attempting to represent variability and uncertainty in inputs for which they have no situation specific information. For some inputs this constitutes an effective use of existing data since the distribution is consistent across a range of locations and populations. For example, adult body weight frequently is represented using information from a general United

---

[1]McKone and Ryan, 1989; Burmaster and von Stackelberg, 1991; Belcher and Travis, 1991; Hawkins, 1991; McKone and Bogen, 1992; Thompson *et al.*, 1992; Helton, 1993; Finley and Paustenbach, 1994; Hoffman and Hammonds, 1994; Hattis and Burmaster, 1994; MacIntosh *et al.*, 1994; Lee *et al.*, 1995; Cullen, 1995; Frey and Rhodes, 1996; Price *et al.*, 1996a&b.

States population survey. This approach is sound except in rare scenarios where the assessment of exposure to a subpopulation which differs from the general population (such as pregnant women) is of interest. For inputs that may vary on a small spatial scale, such as soil concentration of PCBs, it is difficult to imagine accurately representing one scenario with information collected in another.

One of the main messages of this book is that the development of representative distributions should be tailored to their intended use in support of decision-making. This requires a full understanding of the appropriate spatial and temporal averaging, as well as the population and environmental agent of concern. Under this assumption, published distributions from other sites, populations, and scenarios, sometimes may serve as the *starting point* for distribution development in a new analysis.

6.1.1. Suggested Information to Accompany Initial Distributions

We turn now to this possibility and ask: What information should accompany published initial distributions for exposure factors (or sometimes raw data sets) in order to ensure their appropriate adaptation in subsequent analyses? What follows is a wish list for the kind of information that may be useful in this context.

- Who collected the data and for what purpose?
- What was the sampling scheme? What averaging time, spatial coverage, sample size, and population are represented by the data?
- What methods of data collection and analysis were used?
- Was there any censoring or truncation of the measurements for any reason (e.g., limited range of calibration, etc.)?
- What are the strengths and weaknesses of the data in the eyes of those who carried out the collection and analysis?
- What is the range of appropriate uses of the data in the eyes of those who carried out the collection and analysis?
- What related data sets exist?
- What makes this data set unique?
- How was the distribution fit to the data, by whom, and for what purpose?
- Were multiple methods attempted for developing a distribution and did they yield consistent results?
- On what basis were relevant data judged to be representative, or how was the distribution adjusted to account for nonrepresentativeness of the data?
- Was the distribution based solely on elicitation of expert judgment? If so, what are the qualifications of the expert(s), what elicitation protocol was used and what explanations were given for the judgments?

6.2. MAXIMUM ENTROPY INFERENCE APPROACH

Maximum entropy inference has been used in a variety of applications for decades (Shannon and Weaver, 1949; Jaynes, 1957; Levine and Tribus, 1978; Buckley, 1985; Harr, 1987; Kaplan, 1991; Kapur and Kasavan, 1992) and recently has been used in the development of distributions for exposure model inputs. The mathematical theory appears in the early references. We focus only on the results of this research adapting the approach of Lee and Wright (1994) and illustrating with an example.

6.2.1. The Basis of Maximum Entropy Inference

The variability or uncertainty inherent in a distribution may be described by a property known as the entropy of the distribution. Shannon and Weaver (1949) defined entropy, H, for a continuous distribution as:

$$H = -\int p(x) \log_2(p(x))dx \qquad (6.1)$$

where $p(x)$ represents the probability density function associated with the distribution and log_2 is the logarithm in base 2.

Maximum entropy inference allows the analyst to identify the distribution with maximum entropy, which is consistent with available information about an input. This technique produces distributions that are very broad since no mathematical possibility is ignored unless it is specifically precluded by known constraints. In other words, the maximum entropy approach generates a distribution, which is restricted by the information available, (but no further). Also it can help address several potential concerns with informal subjective approaches to probability assessment, such as poor quantification of uncertainty and unspecified assumptions, but does not address others, such as imprecision, uncorrected biases, and lack of credibility.

Consider an unknown input about which we have information concerning one or more of the following properties:

1. the lower bound, a,
2. the upper bound, b,
3. the mean, μ, and
4. the standard deviation, σ.

Working from unbiased estimates of a subset of at least two of these properties, an analyst can develop distributions based on the maximum entropy solutions provided by Shannon and Weaver (1949) and Harr (1987). The following four entropy maximizing solutions for common cases ensure distributions that maximize uncertainty about the input. Answers are also consistent with the requirement that any probability distribution must integrate to 1, (i.e., $\int p(x)dx = 1$).

    a. With estimates of the upper and lower bounds, a uniform distribution is developed, $U[a,b]$.
    b. With estimates of the mean and standard deviation, a normal distribution is defined, $N(\mu,\sigma)$.
    c. With estimates of the lower bound and mean, an Exponential distribution is developed. The rate parameter of the Exponential, λ, is defined: $\lambda = 1/\mu$.
    d. With estimates of the lower bound, upper bound, mean and standard deviation, a beta distribution is defined, $B(\alpha_1,\alpha_2)$ where α_1 and α_2 are defined in Eqs. (6.2) and (6.3), and the beta distribution is defined in Section 4.3.1.5..

Maximum entropy inference produces distributions quickly and inexpensively. Granted these are broad distributions encompassing as much uncertainty as is warranted by the information available. We would caution that the resulting distributions will be biased if biased estimates of the properties discussed (upper bound, lower bound, mean, and standard deviation) are incorporated.

6.2.2. Example

Maximum entropy inference is used to develop a distribution for the fraction of time spent indoors by an individual, Fi, in Chap. 9. For this input we assume that the lower bound (a) is 0.8, the upper bound (b) is 1.0, and the most likely value (μ) is 0.96 (see Section 9.2.4.2 for additional detail).

From Lee and Wright (1994) we find the following formulae derived for fitting the parameters of a beta distribution, $B(\alpha_1, \alpha_2)$, when the lower bound, upper bound and most likely value are available.

$$\alpha_1 = (\mu - a)\left[\frac{\frac{(\mu - a)(b - \mu)}{\sigma^2} - 1}{b - a}\right] \tag{6.2}$$

$$\alpha_2 = \frac{(\mu - a)(b - \mu)}{\sigma^2} - 1 - \alpha_1 \tag{6.3}$$

Eqs. (6.2) and (6.3) assume a beta whose lower bound and upper bound are 0 and 1 respectively. For cases in which the lower bound is not 0, such as ours, the formulae for μ and σ are modified as follows:

$$\mu_T = \frac{\mu - a}{b - a} \tag{6.4}$$

$$\sigma = C_v(\mu - \lambda) \tag{6.5}$$

where:

$$C_v = 0.99 - 0.4\mu_T - 1.15\mu_T^2 + 0.55\mu_T^3 \tag{6.6}$$

Estimation of α_1 and α_2 then proceeds using the original untransformed inputs.

For our example, the solution is a maximum entropy beta distribution to represent fraction of time spent indoors $\beta(3.5, 0.876)$ with scale = 0.2. However remember that the fraction of time spent indoors is assumed to fall between 0.8 and 1.0. Thus, we scale the beta to produce values between 0 and 0.2, which are then back-transformed. Fig. 6.1 shows the probability density function for the beta distribution representing the fraction of time spent indoors. Fig. 6.2 shows the simulated frequency histogram, a discrete image of 1000 random draws from the probability density function. The simulated values were back-transformed, by adding 0.8 to each, to generate the cumulative distribution function (Fig. 6.3).

6.3 . COMBINING INFORMATION

As outlined in Chap. 4, we rarely have abundant, high quality and relevant information for all quantities of interest in an analysis. Thus we may seek to

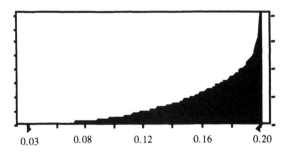

FIGURE 6.1 Probability density function for fraction of time spent indoors (transformed values).

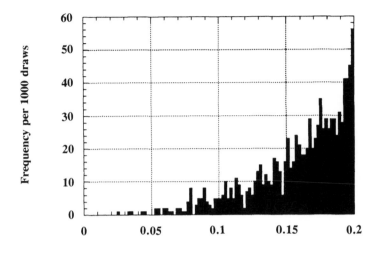

FIGURE 6.2 Simulated histogram for fraction of time spent indoors (untransformed values).

combine information from different sources or of different types to make up for the gaps. On many occasions, the motivation for such efforts is to strengthen our ability to draw conclusions based on the data, such as decisions about the form or parameters of the underlying distribution representing variability and uncertainty in a quantity of interest. However, the combination of information does

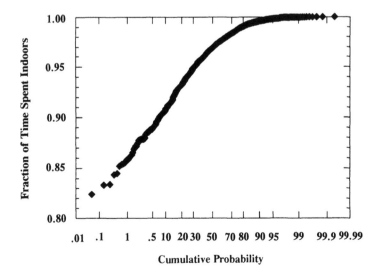

FIGURE 6.3 Cumulative distribution function for fraction of time spent indoors (back-transformed values).

not always successfully serve this goal. For example, it is important to acknowledge that combining information can lead to an apparent increase in uncertainty, especially when one or more of the assumptions underlying one's previous state of knowledge is flawed (Rhodes and Frey, 1997). This result should be viewed as a positive step, because it leads to a more appropriate estimation of the uncertainty in a quantity or context. The American Statistical Association (ASA, 1992) has carried out extensive research regarding what, why, whether, and when to combine information. It is from their excellent work that we draw the following brief discussion.

6.3.1. What to Combine?

There is no unique categorization of the full range of information with which exposure assessors work and which they might want to combine. There exists a spectrum from empirical data to subjective judgment. Depending on the purpose of a particular analysis one might wish to combine information from multiple related sources or from distinct and unrelated sources.

In the simplest case, one could imagine combining information within a single data set. A common example of this is averaging a variable across space or time or a population. For example, in Chap. 9 we wish to combine air concentrations representing variability, averaging across a year. In the case of outdoor (ambient) air concentration depends on temperature, and therefore on season. Thus it may be necessary to group the data by season prior to averaging and then to reweight to account for uneven sampling and to make the data representative of the entire year.

A second case involves the availability of multiple sets of measurements of a common quantity. One interesting example is the exploration of multiple sets of measurements from which a physical constant is estimated, such as the speed of light. Henrion and Fischhoff (1986) note that over time a whole range of values representing the speed of light have been estimated based on various experimental data sets, some of which fall outside of the error bounds for the previously estimated value. Perhaps in this case it is the differences between the information sources, and the handling of error in each, that are most interesting. Rather than combine the individual data sets into a single set from which to estimate the speed of light, it is useful to consider the prediction of the speed of light based on each, and to try to understand the source of differences and of underestimation of uncertainty.

A third case is characterized by the availability of both empirical measurements and subjective judgment. In Chap. 9 we consider a quantity for which the measured values are not directly relevant in the context of interest, i.e., the fraction of time that individuals spend inside their homes, Fi (see Section 9.2.4.2). Because the available data sets pertaining to Fi are of questionable relevance to the site and population of interest, we proceed to develop a distribution based on knowledge of the climate and population in New Bedford, using an informal judgment approach. Upon comparing our judgment based distribution for Fi with the distribution based on the values in the data set, we find that the two types of information lead to a similar "most likely" value for Fi. A more formal approach to combining data and judgment involves the use of Bayes' Theorem, as discussed in Chaps. 2 and 8 (for additional information see March and Simon, 1958; Raiffa and Schlaifer, 1961; DeGroot, 1969).

A fourth case requires the combination of separate sources of information representing central values of an input versus extreme values in the tails of its distribution. This case could result in a mixture model of the type discussed in Chap. 4.

A final case involves combining judgments across separate individuals, or judgments from a single individual across different elicitations. Where there are limited data or alternative theoretical bases for modeling a system, experts may disagree on the descriptions of variability (or uncertainty) when generating frequency (or probability) distributions. In this case it is possible to combine information using Bayes' Theorem, but it also may be informative to explore the implications of using the judgments of different experts in an analysis to determine whether substantially different conclusions result. If the choice of expert, or the choice of approach to combining the judgments of multiple experts, is influential then one should evaluate the source(s) of the difference(s) (Morgan and Henrion, 1990). Recent examples of combining judgments across individuals to generate distributions of cancer potency have been published by Evans *et al.* (1994a&b). In an initial effort the authors used a combined set of their own judgments about the carcinogenic potency of formaldehyde to demonstrate their techniques (Evans *et al.*, 1994a). In a subsequent effort, judgments were elicited from six toxicologists and three epidemiologists and combined to describe the carcinogenic potency of chloroform (Evans *et al.*, 1994b).

6.3.2. Why Combine Information?

Why do analysts combine information? As stated above, multiple sources of information may be used to estimate some quantity of interest, thus combining them may sharpen the estimate of that quantity. This result would be expected in cases where individual sources contain small numbers of measurements which are consistent in central tendency. However, combining sources of information does not always produce answers which are less uncertain than those produced by considering only individual data sets. For this reason there are a number of useful statements that should accompany the results of combined information. First there should be some identification of data sources or sets which contributed most significantly to the final combined result. When data and judgment are combined, it is important to assess qualitatively, and if possible quantitatively, the relative contribution of each type of information to the end result. In addition, an indication of one's level of confidence in the end result can be very valuable.

6.3.3. When to Consider Preserving Separate Sources of Information?

There are deceptively simple rules of thumb about when it is appropriate to combine information, and yet in practice they demand enormous skill. For example, information can be successfully combined when the multiple information sources are similar enough that you will end up with a sensible result. Further, information can be combined if you are able to express your judgments about their similarity in a form that provides the means to combine them.

Guidance about when not to combine may be even more important than guidance about when to combine. Some manipulations reduce the specific content of individual points or sources of information. For example, simple averaging smooths the extremes observed in any raw data set. When the most interesting analytical questions seek to identify the conditions under which extreme or outlying values are observed, averages are not helpful. Further, if only a combined data set is maintained, one is no longer in a position to assess differences in central tendency or variability across data sets. In many cases, it is more informative to look at individual sources of information carefully and separately than to combine. Another case in which combining would not be helpful is in the assessment of interindividual variability in an exposed population or the identification of a highly exposed subpopulation.

6.3.4. An Example

As noted, there are both simple and complex examples of combining data. For example, in Chap. 9 we grapple with developing a distribution for the concentration of PCBs in outdoor air averaged over a year, where we have a set of measurements distributed across the year in a nonrandom fashion (Table 9.5). It has been observed that the concentration of semivolatile compounds in air varies with ambient temperature. Since sampling is more likely to be conducted in fair weather, variability across the year may be underestimated in a simple pooled analysis of the data (in which each data point receives equal weight).

As an initial step toward generating a distribution for average PCB concentration in outdoor air across the year, we develop a distribution of variability in concentration for a single randomly selected day (Section 9.2.2.2). First, we explore the need to reweight the data from each season for the reasons discussed above. For the weighting we establish three seasons according to the ambient temperature conditions in New Bedford: summer (June, July, August), winter (December, January, February), and spring/fall (October, November, March, April, May). We construct seasonal weights according to the fraction of the year that each season is experienced in New Bedford, i.e., summer 33%, winter 25%, and spring/fall 42%.

We assume that the logarithms of the concentration data are distributed normally in each season based on theory presented in Chap. 4 and the case made in Chap. 5 (see Figs. 6.4 to 6.6).

The means and standard deviations of these distributions are calculated individually by season, where an index of 1 is assigned to summer, 2 to winter, and 3 to spring/fall, (i.e., $\mu_{\ln(C_1)}$, $\mu_{\ln(C_2)}$, $\mu_{\ln(C_3)}$, $\sigma_{\ln(C_1)}$, $\sigma_{\ln(C_2)}$, $\sigma_{\ln(C_3)}$, respectively). Then the mean and variance of the distribution of concentration ($\mu_{\ln(C)}$, $\sigma^2_{\ln(C)}$) were constructed using these statistics and the fraction of the year represented by each season ($p_1 = 0.33$, $p_2 = 0.25$, $p_3 = 0.42$, respectively) as follows:

$$\mu_{\ln(C)} = p_1\mu_{\ln(C_1)} + p_2\mu_{\ln(C_2)} + p_3\mu_{\ln(C_3)} = -12.4, \qquad (6.7)$$

$$\sigma^2_{\ln(C)} = p_1(\sigma^2_{\ln(C_1)} + \mu^2_{\ln(C_1)}) + p_2(\sigma^2_{\ln(C_2)} + \mu^2_{\ln(C_2)})$$
$$+ p_3(\sigma^2_{\ln(C_3)} + \mu^2_{\ln(C_3)}) - \mu^2_{\ln(C)} = 2.5 \qquad (6.8)$$

thus, $\sigma_{\ln(C)} = 1.6$, using the following means and standard deviations for the logtransformed data (where the raw data are in units of mg/m³):

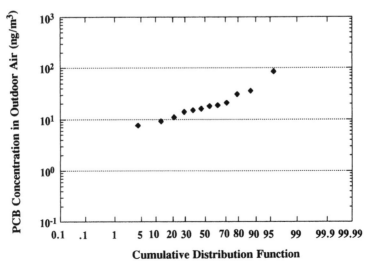

FIGURE 6.4 PCB concentration measurements in outdoor air in summer.

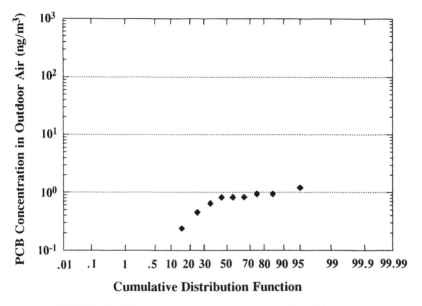

FIGURE 6.5 PCB concentration measurements in outdoor air in winter.

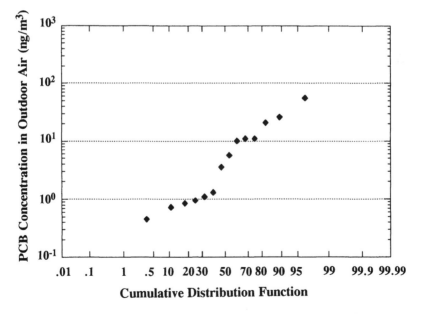

FIGURE 6.6 PCB concentration measurements in outdoor air in spring/fall.

$$\mu_{\ln(C_1)} = -11.0,$$
$$\mu_{\ln(C_2)} = -14.1,$$
$$\mu_{\ln(C_3)} = -12.4,$$
$$\sigma_{\ln(C_1)} = 0.48,$$
$$\sigma_{\ln(C_2)} = 0.31, \text{and}$$
$$\sigma_{\ln(C_3)} = 1.55.$$

In a comparison of the reweighted pooling of the seasonal data sets described above with simple pooling, we find that the mean and standard deviation of the logtransformed data are −12.4 and 1.6 for both approaches. Further, both distributions resemble the distribution of outdoor air concentration in the spring/fall season (see Fig. 6.7). This result is due to the fact that the spring/fall season constitutes the largest fraction of the year and carries the greatest variance in concentration.

6.4. SURPRISE

Overconfidence, a common human trait, causes analysts to experience a great deal of difficulty in the estimation of the magnitude of true uncertainty

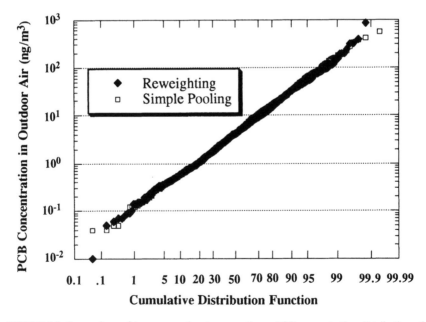

FIGURE 6.7 Comparison of two approaches to generating a PCB concentration distribution of variability in outdoor air.

(Morgan and Henrion, 1990). As might be expected, uncertainty about many exposure inputs also is subject to underestimation. The use of point estimates to represent uncertain quantities is an extreme example of this. One approach, used to combat this problem, is extremely careful and rigorous characterization of both variability and uncertainty. To the discussion presented in Chap. 5, we add a set of tools aimed specifically at addressing surprise, based on the work of Shlyakhter (1994).

6.4.1. Defining Surprise

Many definitions of surprise are possible; however, the term often is used to refer to outcomes which were largely unanticipated (sometimes those with adverse consequences). Probabilistic analysis is well suited to address a particular variety of surprise: the surprise accompanying the realization that the true value of an uncertain quantity is far from someone's "best guess." In the context of elicitation of subjective probability distributions, *surprises* may be defined as true values that fall above the 99th percentile or below the 1st percentile specified by an individual who is queried about an uncertain quantity. The frequency of surprises, that is, the fraction of true values falling outside this 98% credible interval, is sometimes called the *surprise index* (Morgan and Henrion, 1990). One reason for not anticipating extreme values for uncertain quantities is the tendency to place too much credence in observations from small, nonrepresentative, data sets. One may be tempted to use the upper and lower bounds of the data as an estimate of the upper and lower bounds of the quantity of interest, although, higher or lower values might be obtained from a more representative sample. In this chapter we focus on surprise associated with model inputs. Readers will find additional information about model uncertainty in Chap. 3 and uncertainty associated with small, but representative, data sets in Chaps. 5 and 9.

Unsuspected errors have been observed in many types of predictions, including physics, energy modeling, and population forecasting. For example, pioneer chemical power plants have often been subject to unanticipated cost growth or performance shortfalls. The unsuspected errors may be associated with model inputs or model structure itself. In addition, incomplete characterization of pollutant sources, exposure pathways, or exposure scenarios may lead to large errors. For example, after decades of regulatory pressure on industrial sources of benzene emissions and compliance monitoring of these facilities, a personal monitoring study showed that nonindustrial sources of benzene result in significantly higher levels of human exposure than industrial facilities, to the surprise of many regulators (Wallace, 1995). In another example, recent direct measurements of the neutron dose from the atomic explosion in Hiroshima have revealed a surprising inadequacy in the underlying exposure model. Early calculated doses were found to be underestimates of the true dose by a factor of 10 at distances where exposed humans survived. This discrepancy was attributed to the cracks that may have resulted from fire-testing the atomic bomb's gun-type casing with conventional explosives. It is extremely difficult to estimate the fraction of neutrons that could have leaked through these cracks when the weapon ultimately was deployed (Jablon, 1993).

6.4.2. Accounting for Surprise

Even where the functional forms of the probability distributions are perfectly consistent with the underlying processes that generated the data, if the parameters of those distributions are determined from data from a nonrepresentative scenario, population or location they may be biased (see Chap. 4). Further, distributional parameters determined from a small set of data, albeit scenario appropriate, also may be biased and/or may underestimate spread (see Chap. 5).

The commonly used 95th percentile of an output distribution depends on the "length" of the tails of the input distributions and therefore is sensitive to the coefficient of variation of the inputs. By contrast, the specific choice of distributional family may have little influence on estimates of percentiles even as high as the 90th or 95th percentile of the output. For example, the product of three uniform distributions (range 0 to 10, and coefficient of variation, 0.6) and the product of three normals ($\mu = 5$, $\sigma = 3$ and coefficient of variation, 0.6) possess means and 95th percentiles within a few percent of one another (Taylor, 1993). In other words, the dissimilarity of the uniform and normal input distributions would not affect the 95th percentile of an output distribution computed as a product of three such inputs.

Nevertheless, cases may arise in which the choice of a distributional form or parameterization would affect the tail of the output distribution significantly. For example, the assumption of a nonsymmetric distributional form such as a lognormal for a model input, rather than a symmetric form such as the normal or uniform, could lead to significant differences in the tail of the output distribution. In the case of difficult choices between distributional forms to represent inputs, the sensitivity of simulation results to that choice should be established through experimentation and reported.

The history of science shows that experts often overestimate the reliability of their uncertainty estimates for constant quantities. Experts whose judgments tend to be too narrow, in other words not acknowledging enough uncertainty, are referred to as *overconfident*. The result of overconfidence is that the frequency of surprise for these individuals may be higher than expected. Statistical analysis of time trends in sequential measurements of a physical constant, and comparison with values available at a later point in time, provides some guidance about the credibility of uncertainty estimates (Henrion and Fischhoff, 1986). Cumulative distributions of deviations from true values show a general pattern:

long tails that do not follow a normal distribution, but rather can be represented by an exponential whose slope should be determined by the data.

A prudent analyst may hedge against underestimation of the width of the distribution of an important exposure input, using a framework developed by Shlyakhter (1994). The recommended steps are as follows. First, for a normally distributed parameter, let us define a new variable $t = \Delta'/\Delta$ where Δ is the estimated standard deviation, and Δ' is the true but unknown standard deviation. To represent t one uses the right half of a normal distribution:

$$f(t) = 0 \quad \text{for } t \leq 1 \tag{6.9}$$

$$f(t) = \left(\sqrt{\frac{2}{\pi}}\right)\left(\frac{1}{u}\right)\left[\exp\left(-\frac{(t-1)^2}{2u^2}\right)\right] \quad \text{for } t > 1, u > 0 \tag{6.10}$$

The truncated distribution reflects the fact that experts are mostly overconfident about their uncertainty estimates. A new parameter, u, is introduced to allow a means of accounting for this tendency. If $u \ll 1$, only values close to $t = 1$ (i.e., $\Delta = \Delta'$) are important, and the distribution of the parameter under consideration is normal. Analysis of historical data sets suggests u is approximately 1 for physical measurements, and u is approximately 3 for energy and population projections. This approach, in effect, increases the probability weight in the tails of the distribution. Accounting for unsuspected errors by adjustments to variance in distributions is a valid step toward combating overconfidence

As stated in the introduction to this section, there are multiple measures available to combat overconfidence. The most basic protection is a rigorous exploration of all sources of variability and uncertainty and an honest accounting for the gaps and lacks in the information used.

6.5. DEPENDENCE AMONG INPUTS

For simplicity and accuracy when using a conventional probabilistic simulation technique, the model should be structured so that its input variables are as independent as possible. However, one must still identify potential dependencies between input variables. In some cases, it may be necessary to represent such dependencies explicitly in a probabilistic simulation in order to obtain credible results. In other cases, the model outputs may be insensitive to dependencies among inputs.

As an example of dependence, when the concentration of a pollutant is high, the inhalation rate may be low. This can occur if the pollutant has a noticeable odor and if exposed individuals therefore make an effort to reduce their activity levels or breathing rates of air when the odor is present, or if they leave the contaminated area.

Sometimes it is difficult to predict the direction of correlation. For example, when an individual is known to have a high consumption rate of one type of food, it may imply either a low consumption rate for other foods, since the individual may consume that food in place of several others, or it may imply a high consumption rate for other foods as well, as might be expected for a large person.

There are several approaches to addressing dependence. Two of the most common are (see Section 7.5): (1) modeling the source of the dependence explicitly using a more detailed model; and (2) simulating dependence using multivariate distributions or "restricted pairing" techniques for sample generation. The first approach is preferred whenever possible. This is because many of the approaches to accounting for dependencies are at best approximations that are unable to capture the full complexity (e.g., nonmonotonic relationships) that may exist between model inputs.

Even if dependence exists between input variables, it may have little effect on the overall model results (Smith et al., 1992). Weak correlations and correlations between inputs that do not contribute significantly to the variance in the output, can be safely ignored without compromising the analytical results. Smith et al. (1992) found that substantial bias in the 95th percentile of an output distribution can only occur if two variables with similar and substantial variances are highly correlated. When dependent inputs contribute significantly to variability and uncertainty in the model output, the impact of correlations will depend upon the strength of the correlation and upon what percentiles of the output distribution are of interest.

Where a correlation between two input variables is believed to exist, but a definitive estimate is elusive, a sensitivity analysis can help determine whether its existence could affect model results. Alternative assumptions regarding correlations between inputs to a model can be evaluated with respect to their impact on model results. In cases where correlations among inputs significantly change model results (e.g., by increasing the range of outcomes in the output variable), the analyst may wish to construct a more detailed model to explicitly capture the dependence.

There are a variety of methods available for simulating correlation between inputs. In some cases, dependence between two random variables can be characterized or approximated by a correlation coefficient. Many probabilistic simulation computer packages accept user-specified correlation coefficients, which are then used to induce a statistical covariation between the input variables using restricted pairing techniques (e.g., Decisioneering, 1996; Iman and Conover, 1982; Iman and Shortencarier, 1984). There are also methods for simulating Kendall's tau and Spearman's rho rank correlation (Nelson, 1986), and Pearson product moment correlation (Scheuer and Stoller, 1962) as summarized by USEPA (1996b). Another approach is to create groups of parameters that covary, and treat each group as a new parameter (e.g., intake rate per unit body weight, rather than intake rate and body weight treated as separate variables). Alternatively, one could account for correlation with multivariate distributions encompassing the correlated inputs (e.g., bivariate normal). Stratification of the population into relatively homogeneous subgroups minimizes variability within each subgroup and may reduce the importance of possible correlations among inputs. Finally, one could use dependency bounds to characterize upper and lower bounds on an analysis regardless of the underlying correlation structure (e.g., Ferson and Long, 1995).

As an example of the use of multivariate distributions, we consider the characterization of dependencies by resampling from a large dataset (Singh, 1997). In estimating the uncertainty in peak ozone concentrations based upon emissions of ozone precursors from highway vehicles, it is necessary to estimate the hour-by-hour emissions of air pollutants. The uncertainty in emissions is correlated from one hour to the next. For example, if vehicle miles traveled in the hour beginning at 8 AM is high, then vehicle miles traveled in the subsequent hour is likely to be high, and vice versa. Based upon statistical analysis of hour-by-hour vehicle count data obtained from the North Carolina Department of Transportation, it was determined that the correlation between vehicle counts in adjacent hours was typically higher than 0.8. In addition, the correlation between morning and afternoon rush hours was also significant. Since the hourly vehicle counts must sum to a known daily vehicle count, rather than try to model the uncertainty in vehicle counts independently for each hour, the entire data set was resampled. This was done by assigning an equal probability to drawing hourly count data from each of the available days of data in the data set. Thus, each time one day was sampled, an internally consistent set of hourly vehicle counts was obtained. This approach requires no additional assumptions on the part of the analyst regarding either the probability distributions to use for vehicle counts in any given hour nor regarding correlation or dependence between the distributions for different hours.

An alternative to the use of correlations is to employ dispersive Monte Carlo sampling, which is intended to characterize the maximum possible dispersion in a model input based upon linear correlations (Bukowski et al., 1995; Whitt, 1976). This technique also is useful for bounding results when the strength of linear correlations among model inputs is not known.

6.6. EXPERT ELICITATION

As discussed earlier two fundamental approaches are commonly used for encoding uncertainty in terms of probability distributions: statistical estimation and judgment. These methods may be used individually or combined, depending on the abundance and representativeness of available information. In the presence of abundant and randomly sampled abundant data statistical approaches are readily applied. In the face of non-random non-representative or scarce data, the elicitation of subjective judgment, particularly from experts, becomes increasingly important.

Some authors suggest that one has a choice about whether or not to use expert judgment in support of exposure and risk analysis. Evans et al. (1994a&b) suggest that the real choice is not whether or not to use these techniques, but rather where to draw the boundary of analysis. Some analysts draw a "tight" analytic boundary—answering only those questions that can be directly answered by reference to frequentist concepts of probability and leaving questions of relevance or interpretation unanswered. Others draw a "loose" analytic boundary, including questions of relevance and interpretation in the formal analysis. In doing so they rely on both frequentist and subjectivist notions of probability. The "tight" analysis favors "objectivity" at the cost of relevance. The "loose" analysis favors "relevance" at the cost of objectivity.

In this section we present a very brief description of expert elicitation and provide references for readers interested in learning about these techniques.

6.6.1. Sources of Bias in Judgments about Uncertainty

There are a number of human behaviors, observed by psychologists, that can lead to bias in one's judgments about a probability distribution to represent a quantity (Slovic et al., 1979; Kahneman et al., 1982; Morgan and Henrion, 1990). For examples, individuals have a tendency to weight information which comes to mind readily, to be strongly influenced by small unrepresentative sets of data with which they are familiar, to be overconfident and estimate uncertainty too narrowly, to resist changing their minds in the face of new information, to try to influence decisions and outcomes by casting their beliefs in a particular direction, to state their beliefs in a way that favors their own performance or status, to knowingly suppress uncertainty in order to appear knowledgeable, and to persist in stating weakening views simply to remain consistent over time. These behaviors present a challenge to analysts seeking to develop distributions to represent uncertain quantities.

6.6.2. Elicitation Protocols

Most people do not walk around with probability distributions in their heads. Rather, an expert must synthesize information to construct a probability distribution when an analyst asks for one. To counteract heuristics and biases such as those identified above, expert elicitation protocols have been developed for constructing a subjective probability distribution. A very widely reported protocol, the Stanford/SRI protocol, was developed in the 1960s and 1970s at Stanford and the Stanford Research Institute (Spetzler and von Holstein, 1975; Morgan and Henrion, 1990; Morgan et al., 1980; Merkhofer, 1987). Expert elicitation protocols also have been applied extensively at Sandia National Laboratories in the context of quantifying uncertainty in risk associated with nuclear power generation (Hora and Iman, 1989; NCRP, 1996). In addition to these references, we direct readers to a recent and comprehensive book of tools and case studies about the use of expert opinion in scientific inquiry and policy making (Cooke, 1991).

CHAPTER 7

Probabilistic Modeling Techniques

In Chap. 3, we addressed issues regarding uncertainty in the structure of models. We now turn to a different aspect of model uncertainty: how to model the effects of variability and uncertainty in model inputs. Here, we assume the model structure is accurate, and that the only sources of uncertainty are in the inputs. Probabilistic analysis can be used to help evaluate the resulting uncertainty in the model output by providing an estimate of the precision of the outputs that can be compared to any available datasets. Probabilistic analysis is also useful to decision-makers who must identify highly exposed individuals or other populations or subpopulations in an exposure assessment. Furthermore, probabilistic analysis can be used to identify the key sources of variability and uncertainty in model inputs for the purposes of targeting further data collection or research.

There are a variety of ways to propagate information about variability or uncertainty through a model. Although we will focus on the use of numerical simulation methods, we will first briefly introduce some analytical and approximation methods. These methods include results obtained from the central limit theorem, analytical solutions based upon transformation of variables, statistical error propagation, first order methods, and interval analysis. Each of these methods is appropriate in some situations and should be considered along with numerical methods based upon Monte Carlo simulation.

7.1. IMPLICATIONS OF THE CENTRAL LIMIT THEOREM FOR PROPAGATION OF DISTRIBUTIONS

The central limit theorem (CLT) can be stated in a variety of ways. One is the CLT for the sum of independent random variables. The distribution of the sum of independent random variables will approach a normal distribution as the number of random variables becomes large or if no single random variable dominates the sum (DeGroot, 1986). Note that it is not necessary to assume, as long as these other conditions hold, anything about the shape of the distributions for each of the variables in the sum. The only other condition on this theorem is that the third central moments of each variable in the sum must be finite. When these conditions hold, the CLT for the sum of independent variables can be summarized as follows:

$$\mu_s = \sum_{i=1}^{n} \mu_{x,i} \tag{7.1}$$

where μ_s is the mean of the sum and $\mu_{x,i}$ are the means of each of the variables being added together. Also in this case, the variance of the sum is equal to the sum of the variances:

$$(\sigma_S)^2 = \sum_{i=1}^{n} (\sigma_{x,i})^2 \tag{7.2}$$

Similarly, the product of independent random variables approaches a lognormal distribution as the number of random variables in the product becomes large or if no single random variable dominates the product. The latter insight is obtained based upon a simple logarithmic transformation of a product so that it becomes a sum. That is, if we have a product:

$$Y = \prod_{i=1}^{n} X_i \tag{7.3}$$

we can transform it into a sum by taking the logarithm of both sides:

$$\ln(Y) = \sum_{i=1}^{n} \ln(X_i) \tag{7.4}$$

Furthermore, if Y is lognormally distributed, then $\ln(Y)$ is normally distributed, and the CLT for sums applies to Eq. (7.4).

Another statement of the CLT is that if a large random sample is taken from any distribution with a mean of μ and a variance σ^2, regardless of whether the distribution is continuous or discrete, then the sampling distribution of the average of the random sample will be approximately a normal distribution with a mean of μ and a variance of σ^2/n. This is known as the CLT for the sample mean. This theorem is extremely powerful and is useful for estimating confidence intervals on the mean. This theorem also has implications for sampling distributions for many statistics, which can often be approximated by normal distributions as the data set sample size becomes large. It is also useful for distinguishing between uncertainty due to finite sample sizes versus variability in the quantity of interest. We return to this topic in the discussion of propagation of variability and uncertainty in separate dimensions of a probabilistic simulation.

It is important to remember that the implications of the CLT are relevant only if the conditions of the CLT exist for a particular situation. Thus, if you have a model that contains both products and sums, or for which some of the inputs are dominant over others, or for which the inputs are not statistically independent, then the CLT cannot be used to provide insight regarding the exact shape of the output distribution, or of the mean and variance of the result.

7.2. PROPERTIES OF THE MEAN AND VARIANCE

Aside from the CLT, there are other theorems regarding the properties of the mean and variance under various situations. For example, the mean of a sum of random variables for which the means exist is the same as the sum of the means of each random variable. Furthermore, the mean of a constant multiplied by a random variable is the same as the mean of the random variable multiplied by a constant. Thus, the expected value of a linear function of random variables is the same as the linear function of the expected values of each of the model inputs:

$$E\left[\sum_{i=1}^{m} (b_j X_j)\right] = \sum_{i=1}^{m} [b_j E(X_j)] \qquad (7.5)$$

where there are m random variables X_j, and the quantities b_j are constants. Note that there is no requirement that the input random variables be statistically independent in order for Eq. (7.5) to be valid (Ang and Tang, 1975). In addition, there is no requirement regarding the shape of the input random variables. This equation is valid for any correlated or uncorrelated random variables of any distribution. In the case of linear functions with input random variables, these properties are useful for developing analytical solutions for the mean of a function based upon the means of each of the function's inputs.

It also turns out that the mean of a product of *independent* random variables is the product of the means of the variables, as long as the mean exists for each of the inputs (Ang and Tang, 1975; DeGroot, 1986). Here, we use the notation that the mean of a variable X is the expected value $E(X)$:

$$E\left(\prod_{i=1}^{n} X_i\right) = \prod_{i=1}^{n} E(X_i) \tag{7.6}$$

The variance also has some basic properties. For the sum of *independent* random variables, the variance of the sum is equal to the sum of the variances:

$$\text{Var}\left(\sum_{i=1}^{n} X_i\right) = \sum_{i=1}^{n} \text{Var}(X_i) \tag{7.7}$$

Here, we have used the notation that the variance of a random variable X is $\text{Var}(X)$. If each independent input random variable is multiplied by an arbitrary constant and if the sum includes an arbitrary constant, then the variance of the sum is given by (DeGroot, 1986):

$$\text{Var}\left(\left\{\sum_{i=1}^{n} a_i X_i\right\} + b\right) = \sum_{i=1}^{n} a_i^2 \text{Var}(X_i) \tag{7.8}$$

In the case of a product of independent random variables, such as $Y = X_1 X_2 \ldots X_n$, the variance of the product is equal to the product of the variances and an additional term involving the means of each model input:

$$\text{Var}(Y) = \prod_{i=1}^{n} \text{Var}(X_i) - \left[\prod_{i=1}^{n} E(X_i)\right]^2 \tag{7.9}$$

7.3. ANALYTICAL METHODS: TRANSFORMATION OF VARIABLES

Although the results of Section 7.2 are useful in some cases for propagating the mean and variance through a simple linear model, they do not imply anything about the shape of the model output distribution. Thus, if we were interested in making predictions regarding the 95th percentile of the model output for a linear function of independent random variables, we would not have sufficient information based solely on the properties of the mean and variance to do this. However, in some cases, it is possible to develop exact analytical solutions not only for the mean and central moments of the model output, but also for the entire distribution function of the model output. A technique for developing such solutions is known as the "transformation of variables" method. Here, we present some of the key features of this method. More information on this technique is available in Hahn and Shapiro (1967).

Suppose we have a general function:

$$w = h(X), X = \{X_1, X_2, \ldots\} \qquad (7.10)$$

For each of the model inputs, X_i, we specify a marginal probability distribution. Furthermore, we must specify the joint distribution of all model inputs, X. The joint distribution captures any correlations that exist among the marginal distributions. The objective is then to perform a series of steps to obtain an explicit solution for the probability distribution of the model output w as a function of the model inputs, X.

The method of transformation of variables can be used, for example, to show that the sum of two independent normal distributions is also a normal distribution, the mean of the sum is the sum of the means, and the variance of the sum is the sum of the variances. This example can be generalized to the case of the sum of any number of independent normally distributed random variables. In the case of nonindependent normal variables, the sum is also a normal distribution and the mean of the sum is the sum of the means. However, the variance of the sum must be adjusted for correlation among the variables.

The transformation of variables method provides powerful insights regarding the exact shape of the distribution of a model output for cases in which it is possible to perform all of the mathematical steps of which the method is comprised. In cases for which an analytical solution exists regarding the probability distribution for a model output, it is then possible to make predictions for any portion of the output distribution. This is in contrast to the methods presented in Sections 7.1 and 7.2, for which only the mean and variance of the output are known under specific conditions.

It can also be shown that the products and quotients of independent lognormal variables are themselves lognormally distributed. This latter result is perhaps of more practical significance in exposure assessment than is the result for the sums of normal variables. Wilson and Crouch (1981) and Aitcheson and Brown (1957) discuss the properties of equations involving products and quotients of lognormally distributed input variables. Burmaster and Thompson (1995) also illustrate the use of analytical solutions for such equations. The analytical solution for the variance of a model output also can be used to develop measures for identifying sensitive input variables, as discussed in Chap. 8.

In this section, we have discussed the method of transformation of variables and indicated how it can be used to come up with exact solutions for not only the mean and variance of a model output, but also for the exact probability distribution model for the model output. However, this method is limited in application to situations for which mathematical solutions exist. Here, we have described the result for sums of normal variables and for products or quotients of independent lognormal variables. However, if you have a model: (a) which includes both sums and products, such as a multipathway exposure model; (b) for which the inputs are not all of the same distribution as is required for the analytical solution to be valid; or (c) for which any required assumptions regarding independence are not valid, then it may be necessary to consider using other methods for propagating distributions through the models.

7.4. APPROXIMATION METHODS BASED UPON TAYLOR SERIES EXPANSIONS

There are a number of methods based upon the use of Taylor series expansions for propagating the mean and other central moments of random variables through a model. These methods go by many names, including "generation of system moments," "statistical error propagation," "delta method," "first order methods," and others.

The motivation for these methods is that, based upon a sufficient number of central moments for a model output, it may be possible to select a parametric probability distribution model that provides a good representation of the distribution for the output. For example, a specific member of the Pearson family of distributions can be specified based upon the first four central moments (i.e., mean, variance, normalized third central moment—skewness, and normalized fourth central moment—kurtosis). Once a parametric distribution for the output has been selected, then predictions can be made regarding any percentile of the model output. Thus, instead of propagating entire probability distributions for each input through the model, it may only be necessary to propagate the moments of the distributions. Of course, a drawback of this method is that only information regarding the central moments is considered. Information regarding the tails of each input distribution, for example, is not specifically considered. Therefore, the selection of a probability distribution based upon the moments of the model output may not properly capture effects at the tails of the distribution, although it may be adequate for characterization of the central tendencies.

The basic approach is to take a general function, such as:

$$y = h(x_1, x_2, \ldots, x_n) \tag{7.11}$$

and then to expand the function about the point $[E(x_1), E(x_2), \ldots E(x_n)]$ using a multivariate Taylor series expansion. The series is typically truncated at a specified set of higher order terms. For example, if all of the inputs to the function are statistically independent, then the mean of the function, $E(y)$, is approximated by the following Taylor series expansion (Hahn and Shapiro, 1967):

$$E(y) = h[E(x_1), E(x_2), \ldots, E(x_n)] + \frac{1}{2} \sum_{i=1}^{n} \frac{\partial^2 \overline{h}}{\partial x_i^2} \mathrm{Var}(x_i) + \begin{pmatrix} \text{higher} \\ \text{order} \\ \text{terms} \end{pmatrix} \quad (7.12)$$

The function h and the partial derivatives on the right-hand side are evaluated at the point $[E(x_1), E(x_2), \ldots E(x_n)]$. To evaluate the expected value of y using Eq. (7.12) requires one evaluation of the original function, evaluation of the variance for each random variable input to the model, and calculation and evaluation of the second partial derivative of the function with respect to each random variable input. For large models, an analytical solution to the second partial derivative may not be possible or practical. Therefore, a numerical method may be required, which typically would require evaluations of the function at and near the point $[E(x_1), E(x_2), \ldots E(x_n)]$. Thus, the number of computations required to estimate the mean of y is proportional to the number of input random variables.

The variance of the function, $\mathrm{Var}(y)$, of statistically independent random variables is approximated by the following:

$$\mathrm{Var}(y) = \sum_{i=1}^{n} \left(\frac{\partial \overline{h}}{\partial x_i} \right)^2 \mathrm{Var}(x_1) + \sum_{i=1}^{n} \left(\frac{\partial \overline{h}}{\partial x_i} \right) \left(\frac{\partial^2 \overline{h}}{\partial x_i^2} \right) \mu_3(x_i) + \begin{pmatrix} \text{higher} \\ \text{order} \\ \text{terms} \end{pmatrix} \quad (7.13)$$

In order to estimate the variance of y, it is necessary to have estimates of the third central moment of each input random variable. It is also necessary to take the first and second partial derivatives of the function with respect to each random variable. Similar to the estimation of the mean of y, the number of calculations required to estimate the variance of y is proportional to the number of input random variables.

The Taylor series expansion-based methods will fail for locally non-differentiable functions. The issue of where to truncate the series is a significant one for highly nonlinear functions. For situations involving highly nonlinear functions, input variables with large coefficients of variation, or highly skewed inputs, series expansions that are truncated too soon can give misleading results. Furthermore, if correlations among the inputs exist, more complicated series expansions must be used.

To illustrate the use of the approximation methods based upon the Taylor series expansions, we will consider several generic functions as given in Table

7.1. For each of these, Eq. (7.12) and (7.13) were used to estimate the mean and variance, respectively, of the function result. The estimated system moments were compared with the results of a numerical simulation of the propagation of the input distributions using Latin Hypercube sampling (LHS) with a sample size of 32,000. Numerical simulation methods are discussed in more detail in Section 7.5. Here, we assume that, when the two methods disagree, the numerical simulation results are a better representation of the "true" solution than the approximation methods. In all cases, the input variables were assumed to be statistically independent, and each was assigned a mean of one and a standard deviation of one. For the numerical simulations, it was also necessary to assume a distribution shape for each of the inputs. For this purpose, a normal distribution was assumed for all inputs.

It is expected that the Taylor series expansion method should provide an exact solution for linear functions. As expected, for the linear functions shown in Table 7.1 the method of system moments provided the same results for both the mean and standard deviation as did the numerical simulation. As the functions deviate from linearity, we expect to begin to see some shortcomings of the truncated series for the mean and variance. For the three functions involving simple products or quotients, the mean value from Eq. (7.12) was the same as the simulated results. However, the standard deviation from Eq. (7.13) was low compared to the simulation result, and it was not possible to obtain a valid result for the two quotient cases because the solution for the variance was a negative number. In the latter case, it appears that additional terms would be needed in the Taylor series expansion to get a valid result. For the case involving the product of the squares of two random variables, neither the mean nor the vari-

TABLE 7.1 Comparison of the Approximation Method for Propagating System Moments with Results Obtained from Numerical Simulations[a]

Function, Y	System Moments		Simulation	
	Mean	Standard Deviation	Mean	Standard Deviation
$X_1 + X_2$	2	1.4	2	1.4
$X_1 - X_2$	0	1.4	0	1.4
$X_1 \cdot X_2$	1	1.4	1	1.8
$X_1^2 \cdot X_2^2$	3	6.1	3.8	30
$1/X$	2	n/a	2	1.9
X_1/X_2	2	n/a	2	3.5
$\exp(-X)$	0.55	n/a	0.47	0.24
$\exp(X)$	4.1	5.8	91,000	1.60E+07

[a]For all inputs, the mean and standard deviation were both assumed to be unity, and the inputs were assumed to be statistically independent. For the simulation results, Latin Hypercube sampling was used with a sample size of 32,000.

ance from the approximation equations appear to be accurate. Similarly, in the case of the exponential models, neither the mean nor the variance were accurate. These results illustrate that for increasingly nonlinear functions, the Taylor series expansion approach, if truncated after the second terms in the series, appears to fail. Thus, a larger number of terms in the series would be required, which in turn would increase the computational effort.

To further evaluate the approximation methods, a number of computational experiments were performed. For example, consider the function:

$$Y = X^n \qquad (7.14)$$

Several cases were evaluated in which the input random variable was normally distributed with a range of coefficients of variation, and in which the exponent varied from 0.5 to 2. A comparison of results for the mean value of Y for these cases based upon the approximation method and the numerical simulation is shown in Fig. 7.1. In fact, it is difficult to distinguish among the cases because all provide nearly the same results. These results indicate that the approximation method provides a good estimate of the mean of this function under the range of conditions evaluated. However, if the input variable is lognormally distributed, with the same means and variances as in the normally distributed cases, the insights obtained are significantly different, as shown in Fig. 7.2. For the cases with low coefficient of variation, the two solution methods agree. However, as the coefficient of variation increases, and as the model departs from linearity, the solutions diverge. These results imply that the Taylor series expansions used here are inadequate for dealing with highly nonlinear functions with inputs that have skewed probability distributions. In these cases, the evaluation of the expansion at a local point centered at the mean of the inputs does not adequately capture the effect of nonlinearity and asymmetry in the model inputs on the mean of the function.

Similar insights were obtained regarding the estimated variances for the cases outlined in the preceding paragraph. These cases are displayed graphically in Figs. 7.3 and 7.4 for the normally distributed and lognormally distributed inputs, respectively. In these cases, note that the comparison could not be done for exponents other than one or two for the normally distributed inputs with coefficients of variation greater than 0.3. This is because for these situations, a significant number of negative values are sampled in the numerical simulation, leading to infeasible solutions. For the normally distributed inputs, the approximation and simulation results agree within 5% when the input coefficient of variation was 0.3 or less. However, for an exponent of two and higher coefficients of variation, the results disagree. Similar results were obtained for the cases with lognormally distributed inputs, as shown in Fig. 7.4, although the behavior of the

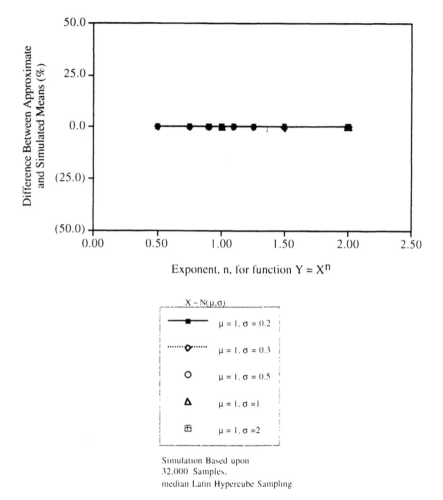

FIGURE 7.1 Evaluation of differences between approximate and numerical solution methods for the mean of an exponential function, $Y = X^n$ with a normally distributed input.

approximation solutions is more complex than for the normally distributed cases. Because the two parameter lognormal distribution is always non-negative, it was always possible to obtain a simulation result over the range of exponents considered.

The effect of correlations among input variables on the solutions for the mean and variance is illustrated using case studies for the following function:

$$Y = X_1 X_2 \qquad (7.15)$$

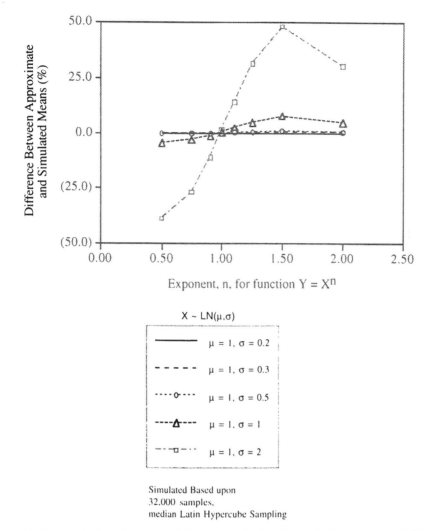

FIGURE 7.2 Evaluation of differences between approximate and numerical solution methods for the mean of an exponential function, $Y = X^n$ with a lognormally distributed input.

Three solutions are considered in cases for which the two input random variables are assumed to be independent. The three cases considered are an exact solution, the Taylor series approximations of Eqs. (7.12) and (7.13), and numerical simulation. The exact solution is based upon Eq. (7.6) for the mean and Eq. (7.9) for the variance. For comparison, a numerical simulation in which the two random inputs have a correlation of $\rho = 1$ is also considered. In all cases, the input random variables were assumed to be normally distributed with a mean

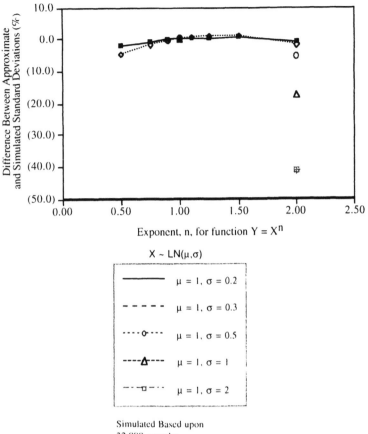

FIGURE 7.3 Evaluation of differences between approximate and numerical solution methods for the standard deviation of an exponential function, $Y = X^n$ with a normally distributed input.

of one. The coefficient of variation was allowed to vary from 0.1 to 2.0 for both inputs.

For the mean of the multiplication function, the exact, approximate, and simulated results were the same for all coefficients of variation considered. For comparison, a simulation of correlated inputs was done. The results are shown in Fig. 7.5. For low coefficients of variation, the correlation among the inputs has little effect on the function result compared to the independent cases. However, as the coefficient of variation of the inputs increases, the mean of the function of the positively correlated inputs becomes larger. Conversely, this example illustrates that even strong correlations may have little effect on the function output if the range of variation of the correlated inputs is relatively small.

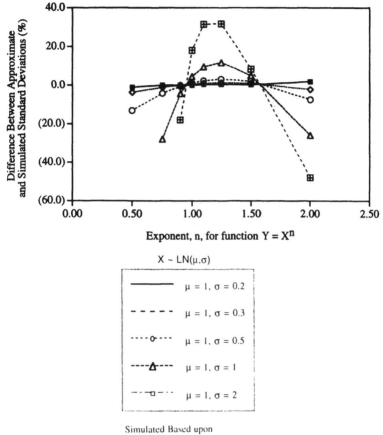

FIGURE 7.4 Evaluation of differences between approximate and numerical solution methods for the standard deviation of an exponential function, $Y = X^n$ with a lognormally distributed input.

In Fig. 7.6, a comparison of solutions is shown for the standard deviation of the multiplicative function of two normally distributed inputs. For the case of independent inputs, the approximate solution underestimates the standard deviation as the coefficient of variation of both inputs becomes larger than approximately 0.5. In contrast, the exact and the simulation results agree well for all ranges of coefficient of variation considered here. For comparison, the simulation of positively correlated inputs provides larger estimates of the standard deviation of the function. However, similar to the results for the mean, the difference between the variances of the result for the dependent and independent cases is small for input random variable coefficients of variation of 0.3 or less.

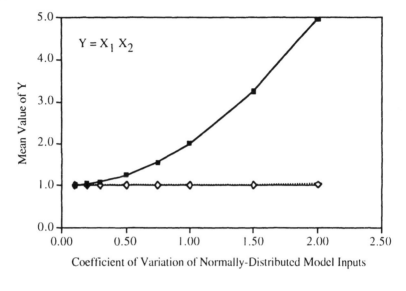

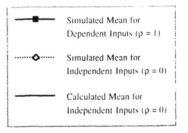

FIGURE 7.5 Comparison of solutions for the mean of a function $Y = X_1 X_2$ based upon Taylor Series expansion and approximate solutions.

7.5. NUMERICAL SIMULATION METHODS

Simulation is a "process of replicating the real world based upon a set of assumptions and conceived models of reality" (Ang and Tang, 1984). Simulations may be either experimental or theoretical. Experimental simulations of exposures and risks can be either costly or unethical, thus numerical simulations are often preferred. One set of methods for simulating the propagation of probability distribution in model inputs through a model is based upon simulated random sampling. The most well-known of these methods is simple Monte Carlo simulation.

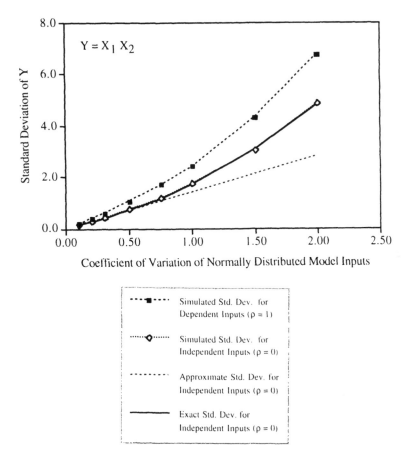

FIGURE 7.6 Comparison of solutions for the standard deviation of a function $Y = X_1 X_2$ based upon exact, Taylor Series expansion and simulation solutions.

As introduced in Chap. 1, Monte Carlo simulation has been around for 50 years, and was first used as a method for solving mathematical integration problems. Some newcomers to probabilistic analysis use the term "Monte Carlo" broadly; however, there are other numerical simulation techniques that are related to Monte Carlo simulation, such as LHS and importance sampling. In addition, there are some techniques for propagating distributions through models that are based on transformations and are not based upon Monte Carlo simulation. We find it convenient to refer to all methods for propagating distributions, or the moments of distributions, through models as "probabilistic techniques," and to categorize Monte Carlo simulation as one type of numerical simulation method. To help clarify the terminology, a taxonomy of commonly used methods for propagating moments or distributions through models is summarized in

TABLE 7.2   A Taxonomy of Common
Probabilistic Methods for Propagating
Moments or Distributions through Models

Analytical Solutions for Moments
  Central Limit Theorems
  Properties of the Mean and Variance
Analytical Solutions for Distributions
  Transformation of Variables
Approximation Methods for Moments
  First-Order Methods
  Taylor Series Expansions
Numerical Methods
  Monte Carlo simulation
  Latin Hypercube Sampling
  Importance Sampling
  Fourier Amplitude Sensitivity Test
  Others

Table 7.2. Another way to view Monte Carlo simulation is by analogy with other types of numerical methods, such as those used for solving systems for differential equations. Monte Carlo simulation is one numerical technique for propagation of distributions through models, just as fourth order Runge Kutta is one method for numerical solution of a system of differential equations (Hornbeck, 1975). Each has its own set of strengths and limitations. However, the specific limitations of Monte Carlo simulation should not be assumed to apply to all numerical methods for propagating distributions through models, just as the specific limitations of fourth order Runge Kutta methods should not be assumed to apply to all numerical methods for solving systems of differential equations.

We will focus our discussion on two numerical simulation methods: Monte Carlo and Latin Hypercube sampling. These are the most commonly used methods. We will briefly mention some other methods, including importance sampling.

7.5.1. Monte Carlo Simulation

In Monte Carlo simulation, a model is run repeatedly, using different values for each of the uncertain input parameters each time (e.g., Hahn and Shapiro, 1967; Ang and Tang, 1984; Morgan and Henrion, 1990). The values of each of the uncertain input parameters are generated based on the probability distribution for the parameter. If there are two or more uncertain input parameters, one value from each is sampled simultaneously in each repetition in the simulation. With many input variables, one can envision Monte Carlo simulation as providing a random sampling from a space of m dimensions, where m is the number of random variables that are inputs to a model. Similarly, one can envision Monte Carlo as providing a simulated set of sample values for the joint distribution of all of the random variable inputs to the model. Over the course of a simulation, perhaps 100, 1000, 5000, or even more sets of samples of the model inputs, and corresponding repetitions of the evaluation of the model, may be made. The result, then, is a set of sample values for each of the model output variables, which can be treated statistically as if they were an experimentally or empirically observed set of data.

Fig. 7.7 illustrates with a flowchart the general process of applying Monte Carlo simulation to a model. For each input to the model which is a random variable, a probability distribution is specified. Random samples are simulated for each of the input distributions. One sample from each input distribution is selected, and the set of samples (sometimes referred to as an "m-tuple," where m is the number of input random variables) is entered into the model. The model is then executed as it would be for any deterministic analysis. The model results are stored. The process is repeated until the specified number of model iterations have been completed. Using Monte Carlo techniques, it is therefore possible to represent uncertainty in the output of a model by generating sample values for the model inputs, and running the model repetitively. Instead of obtaining a single number for model outputs as in a deterministic simulation, a set of samples is obtained. These can be represented as CDFs and summarized using typical statistics such as mean and variance.

Another way to view this is that Monte Carlo simulation enables one to evaluate the integral of a PDF. In fact, it is typically more meaningful to report the results of probabilistic analyses in the form of CDFs. The estimates of the CDF, which is an integral of the PDF, are numerically more stable than estimates of probability density. Furthermore, the CDF allows for quantitative insight regarding the percentiles of the distribution.

Although the generation of sample values for model input parameters is probabilistic, the execution of the model for a given set of samples in a repetition is deterministic. The advantage of Monte Carlo methods, however, is that these deterministic simulations are repeated in a manner that yields important insights into the sensitivity of the model to variations in the input parameters, as well as into the likelihood of obtaining any particular outcome. Monte Carlo methods also allow the modeler to use any type of probability distribution for which values can be generated on a computer, rather than restricted to forms which are analytically tractable.

Most numerical simulation methods, including random Monte Carlo, require the generation of uniformly distributed random numbers between 0 and 1. Given a uniformly distributed random variable, several methods exist from which to

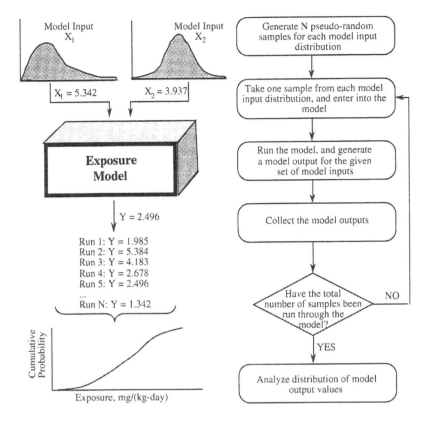

FIGURE 7.7 Schematic and flowchart illustrating the application of Monte Carlo analysis to a model.

simulate random variables that are described by other probability distributions (e.g., normal, lognormal, gamma, Weibull, beta, triangular, etc.). These methods include the inverse transform, composition, and function of random variables (e.g., Ang and Tang, 1984). In addition, methods exist for simulation of jointly distributed random variables, which enables one to represent correlations between two or more simulated random variables.

7.5.1.1. Pseudorandom Number Generator. Simulation methods for dealing with uncertainty are often based upon the use of a random number generator. The random number generator used should come as close as possible to the ideal of generating a series of truly independent random numbers. All random number generators, in reality, will cycle through the same series of random numbers. The key is to make the cycle as long as possible, such that for simulations with large sample sizes and large numbers of probabilistic input assumptions, the random numbers do not contain any significant autocorrelation. While in the past some probabilistic simulation packages have had poor random number generators, in most cases the random number generators underlying such packages on today's market are reasonably good. However, it does not hurt to test the random number generator.

Random number generators are in fact "pseudorandom" number generators, because they are intended to provide *reproducible* sets of numbers which have desirable properties of statistical independence (randomness) and uniformity. Pseudorandom number generators (PRNG) typically have a "random seed" or "starting value" as an input. By changing the seed, one can change the sequence of random numbers obtained. However, if the same seed is used in two or more analyses, then one would obtain the same set and sequence of random numbers. Pseudorandom number generators are typically evaluated using a serial test for independence (randomness) and a goodness-of-fit test for uniformity of the numbers. Barry (1996) describes several evaluation methods, and applies these to a number of commonly used pseudorandom number generators. Barry notes there is no one test that can definitively certify or anticipate the statistical adequacy of every random number sequence that could be produced from a generator. Barry tested a number of pseudorandom number generators and found most did not have any systematic, gross deficiencies. However, there were some common deficiencies. Most PRNGs are based upon "linear congruential methods." These types of PRNGs typically have a period over which a complete sequence of random numbers are produced, and then the same sequence is repeated. Within a cycle of random numbers, there may be auto-correlations and lack of true randomness. To complicate matters, it is possible that particular subsequences of random numbers may have poor behavior, while the overall sequence may appear to satisfy criteria for randomness and uniformity. Among other things, Barry recommends Monte Carlo analyses be repeated with nonoverlapping sequences of random numbers to check for solution stability and repeatability.

7.5.1.2. Simulating Probability Distributions. Once an acceptable series of uniformly distributed pseudorandom numbers has been generated, it is then possible to simulate pseudorandom numbers from other types of probability distributions using various methods. To illustrate how the concepts of Monte Carlo simulation works, it is easiest to consider the inverse transform method. For any probability distribution, it is possible to construct a CDF. As illustrated in Fig. 7.8, the inverse CDF has an abscissa with values ranging from zero to one, and an ordinate with values representing possible outcomes for the random variable of interest. Thus, uniformly distributed random numbers may be used to represent the percentile of the random variable for which a sample is to be generated. The sample values for the random variables are calculated using the inverse CDF functions based on the randomly generated fractiles.

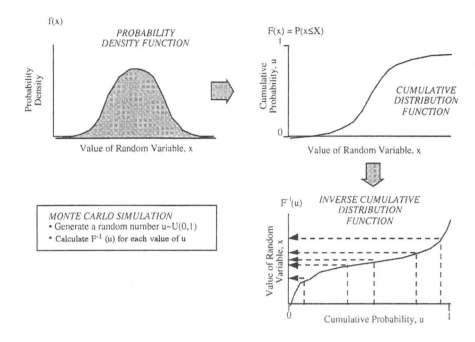

FIGURE 7.8 Monte Carlo simulation (Source: Frey, 1992).

In some cases, random numbers are generated using other methods. In the composition method, the PDF of a random variable is expressed as the weighted sum of other density functions. In the "function of random variables" method, samples from a distribution are simulated using functions of other distributions. For example, a normal distribution may be simulated based upon a function of uniform distributions (Ang and Tang, 1984). For those who would like to write their own computer code for performing Monte Carlo simulation, several texts contain useful information regarding how to simulate random numbers from various types of probability distributions. For example, Hahn and Shapiro (1967) summarize methods for generating exponential, gamma, chi-square (a special case of the gamma), lognormal, Johnson S_B (bounded), Johnson S_U (unbounded), beta, Weibull, Poisson, and binomial distributions. Morgan and Henrion (1990) and Law and Kelton (1991) provide information regarding a similar list of distributions.

7.5.1.3. Simulating Mixture Distributions. In some situations, it may be inappropriate to assign a single parametric distribution to represent variation in a random variable. If a quantity is multimodal, or is the result of different processes that may act at different times or locations, then it may be necessary to represent the quantity with a mixture of distributions.

For example, consider a case related to emissions characterization based upon a paper by Frey and Rhodes (1996). The emissions from a power plant are proportional to the load at which the plant runs over a given amount of time. However, there are some time periods in which the plant is shut down for maintenance, some times during which the plant is changing loads from zero to a minimum "must run" level, and other times during which the plant is running at stable high loads above the "must run" level. These three situations are all components of the overall distribution of plant loads over time. However, there is no single parametric distribution that can properly capture these three types of situations. Therefore, it is necessary to consider a mixture of continuous distributions to represent the plant conditions at other than zero load, and a mixture of discrete and continuous distributions to distinguish between the fraction of time spent at zero load versus the fraction of time spent at non-zero loads.

A data set for plant capacity utilization is shown in Fig. 7.9. These data represent the average plant load over a three-day averaging period, which was of interest in a particular study (Frey and Rhodes, 1996). Based upon a trial and error approach, it was found that the following mixture of distributions would provide an adequate representation of the data set:

$$c_f = \begin{cases} 0 \text{ for } u < 0.08 \\ 0.598 \ \beta(0.475, 0.291) \text{ for } 0.08 \le u < 0.23 \\ N(0.722, 0.074) \text{ for } u \ge 0.23 \end{cases}$$ (7.16)

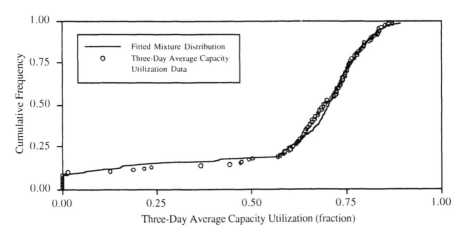

FIGURE 7.9 An example of a mixture distribution (Source: Frey and Rhodes, 1996).

The mixture includes a discrete value of zero for eight percent of the outcomes, and a mixture of two continuous distributions for the remaining 92% of the outcomes. A beta distribution is used to represent the 15% of the outcomes that vary from zero to approximately 0.60 of full load, and a normal distribution was found to adequately represent the remaining 77% of the outcomes.

To implement a mixture distribution in a Monte Carlo simulation requires some careful programming. For a given sample from the mixture, it is necessary to select a component of the mixture, and then draw one sample from that component. Thus, in the example, a uniformly distributed random number, U, from zero to one can be used to make a decision regarding which component will be sampled. If U has a value less than 0.08, then a value of zero will be used. If U has a value between 0.08 and 0.23, then a sample will be drawn from the beta distribution. If U has a value of greater than 0.23, then a sample will be drawn from the normal distribution.

7.5.1.4. Simulation of Correlations. A common but unfounded criticism leveled at Monte Carlo simulation, typically by those unfamiliar with the richness of methods available, is that "Monte Carlo assumes independence" or, similarly, that "Monte Carlo ignores correlations." In fact, it is possible to simulate jointly distributed random variables in which correlations may exist. Smith et al. (1992) propose some rules of thumb for assessing the impact on the model output distribution of including or neglecting correlations among the inputs. There are some limitations specific to simple Monte Carlo in this regard—it is only possible to do this for some special cases. However, there are more general methods for simulating correlations, called restricted pairing techniques, which can be used in conjunction with Monte Carlo methods.

The special case for dealing with correlations in the context of simple Monte Carlo simulation is multivariate normal distributions. For simplicity, we will focus on the bivariate normal distribution. If we have two normally distributed random variables, X and Y, we can represent these as a bivariate normal distribution. The bivariate normal distribution can be used to represent cases in which X and Y are statistically independent, with a correlation of zero, or cases involving negative or positive correlation. As an example, we will consider a case in which the marginal distributions for X and Y are both normal, have a mean of 10 and a standard deviation of 2. We will vary the correlation between these distributions from zero to 0.75.

If we take the marginal distribution of X to be normally distributed with a mean of 10 and a standard deviation of 2, then the conditional distribution for Y has a mean (expected value) of:

$$E(Y|x) = \mu_Y + \rho \frac{\sigma_Y}{\sigma_X}(x - \mu_Y) \qquad (7.17)$$

and a standard deviation of:

$$\mu_{Y|x} = \mu_Y \sqrt{1 - \rho^2} \qquad (7.18)$$

If the correlation, ρ, is zero, then the conditional distribution for Y is the same as the marginal distribution for Y. If the correlation is not zero, then the distribution of Y conditioned on a given value of X will differ from the overall marginal distribution of Y. The equations for the conditional mean and standard deviation of Y can be used to simulate the conditional distribution for Y.

The results of several simulations of bivariate normal distributions are shown in Fig. 7.10. With a correlation of zero, there appears to be random scatter in the sample values for X and Y. As the correlation becomes larger, a more defined pattern emerges. With a correlation of 0.75, it is apparent that the conditional distribution for Y for any given value of X is considerably narrower than is the overall marginal distribution of Y. It is possible to simulate a correlation of $+1$ or -1. In such cases, the data for Y versus X would plot as a straight line with positive and negative slope, respectively.

More generally, correlations among any type and any number of marginal distributions can be simulated using an approximation based upon rank correlation. Rank correlation refers to correlation in the rank ordering of values between two random variables. Thus, rank correlation is a measure of the strength of the monotonic relationship between two variables. A general method for inducing rank correlations between random variables, which is referred to as "restricted pairing," was developed by Iman and Conover (1982). This method was originally developed for use with LHS, which is described in Section 7.5.2. However, it can be used with other simulation methods, such as Monte Carlo, as part of a two step process. In this process, independent marginal distributions are simulated in the first step. In the second step, the ordering of the samples is changed to comply with the user-specified rank correlations among the random variables. To illustrate the use of the restricted pairing method, an example from Frey (1992) is given in Fig. 7.11 using a software package developed by Iman and Shortencarier (1984). Here, a comparison is shown for 100 joint samples drawn from two triangular distributions using LHS. In the first case, statistical independence is assumed. In the second case, a negative correlation is assumed.

A shortcoming of inducing rank correlations is that it can be difficult to simulate a desired sample correlation. Capabilities for simulating correlated random variables exist in some commercially available packages, such as Crystal Ball. Users should verify that the simulations in fact are representative of the desired correlations among random variables. If an unrealistic set of rank correlations are specified, such as A is negatively correlated with both B and C, and

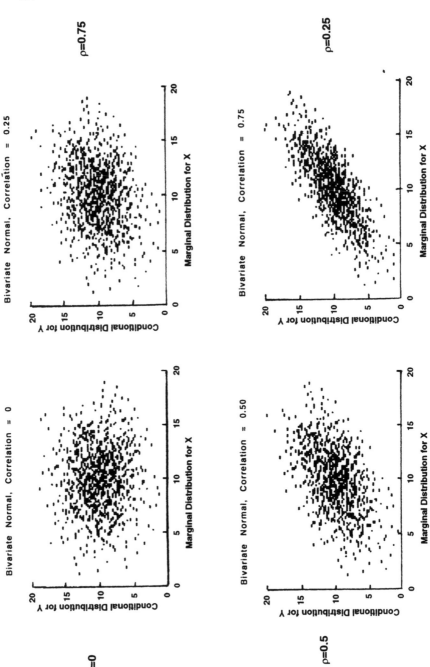

FIGURE 7.10 Simulation of bivariate normal distributions with various correlations among two marginal distributions with mean = 10, variance = 2, using Monte Carlo simulation with sample size = 1000.

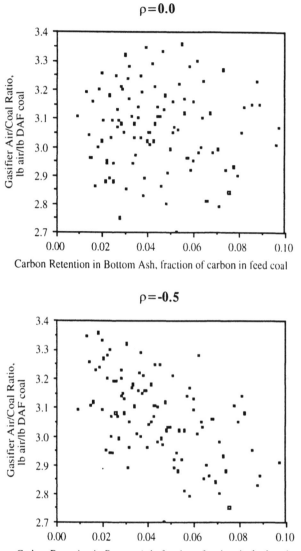

FIGURE 7.11 Simulation of independent and correlated triangular-distributed random variables using the restricted pairing method of Iman and Conover (1982).

B and C are negatively correlated with each other, then meaningless results may be obtained.

It is important to keep in mind the simulation sample size when deciding upon what correlations should be simulated in a given case study. This is because,

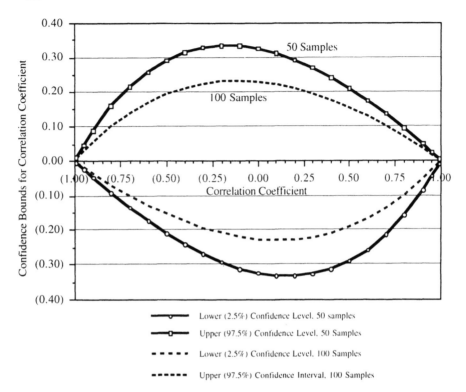

FIGURE 7.12 Confidence intervals on correlation coefficients, based upon the inverse Fisher transformation. The confidence bounds should be added to the nominal value of the correlation coefficient to obtain the confidence interval.

as for any statistic, there is a confidence interval for correlation coefficients that is a function of sample size. The confidence interval for a given correlation coefficient and a given sample size can be estimated using the inverse Fisher transformation method (Steel and Torrie, 1980). An example of the confidence intervals for correlation coefficients, based upon sample sizes of 50 and 100, are shown in Fig. 7.12. The figure illustrates that the confidence interval on a zero correlation can be quite wide for small sample sizes. For example, with only 50 pairs of values for two distributions, a 95% confidence interval for a zero correlation would range from –0.32 to +0.32. Thus, if one were to use random Monte Carlo simulation with a sample size of only 50 to simulate two independent random variables, one would expect that if the simulation were repeated a very large number of times that, on average, 95% of the simulations would yield correlation coefficients somewhere within the 95% confidence intervals. Another way to interpret the practical significance of the confidence interval on the correlation coefficient is that, for small sample sizes, it may not be worthwhile to try to simulate weak correlations. For example, it would be difficult to simulate correlation coefficients of between approximately -0.25 and $+0.25$ with statistical significance with a sample size of only 50. This is because the 95% confidence interval on correlation coefficients within this range for a sample size of 50 would include a value of zero.

The use of common assumptions in different case studies leads to another source of correlations when the case studies are compared. For example, suppose we are comparing two exposure reduction strategies that use similar assumptions regarding the transport and fate of pollutants and human activity patterns, but differ only with regard to the total amount of a pollutant released from an emission source. We could represent the two cases with the following mathematical notation, borrowing from Ang and Tang (1984): let X be a vector of random variables that are common to both control strategies; let A be a set of design variable values specific to Strategy A; and let B be a set of design variable values specific to strategy B. Let us suppose that the output of interest is E_A and E_B, which may represent the distribution of exposures across all members of the population. E_A and E_B are calculated as follows:

$$E_A = f(A,X) \tag{7.19}$$

$$E_B = f(B,X) \tag{7.20}$$

If the inputs X are common to both case studies, then correlated sampling should be used to compare the two strategies. Correlated sampling involves using the same set of sample values for the variables X that are common to both models or case studies. If the same set of sample values are used in both case studies, then any dependence between the two will be captured. For example, if there is a positive correlation between E_A and E_B, then the difference between the two will be less uncertain than if there is no correlation or a negative correlation. As shown in Fig. 7.13, if there is perfect positive correlation ($\rho = +1$), then the difference in exposure between two case studies with same marginal distributions will be zero. However, as the correlation decreases, the differences between the two cases become more uncertain, with an extreme range of uncertainty regarding the comparison for a perfect negative correlation ($\rho = +1$).

The approach described here for accounting for similar inputs to different case studies or models has been used, for example, to compare the performance, emissions, and cost of alternative power generation and environmental control technologies (e.g., Frey and Rubin, 1991; 1992a, 1992b).

7.5.2. Latin Hypercube Sampling

An alternative to random Monte Carlo simulation is Latin Hypercube Sampling (LHS) (McKay et al., 1979). In LHS methods, the percentiles that are used

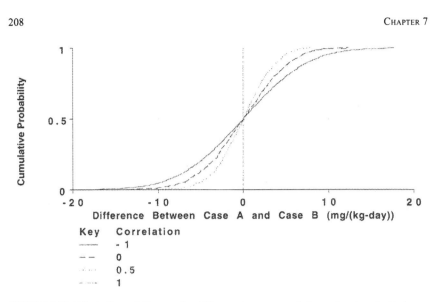

Key   Correlation
‾‾‾‾‾   - 1
‒ ‒   0
· · · · ·   0.5
‒ ·· ‒   1

FIGURE 7.13 Effect of correlations on the difference in exposures between two hypothetical case studies.

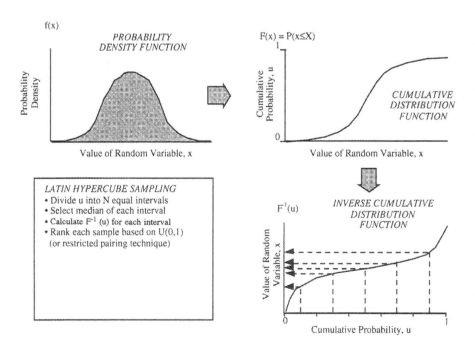

FIGURE 7.14 Median Latin Hypercube Sampling (Source: Frey, 1992).

as inputs to the inverse CDF are not randomly generated. Instead, the probability distribution for the random variable of interest is first divided into ranges of equal probability, and one sample is taken from each equal probability range. This is illustrated in Fig. 7.14. However, the ranking (order) of the samples is typically random over the course of the simulation, and the pairing of samples between two or more random input variables is usually treated as independent. In median LHS, one sample is taken from the median of each equal-probability interval, while in random LHS one sample is taken at random within each interval (Morgan and Henrion, 1990). Random LHS is commonly available in many software packages, such as Crystal Ball (Decisioneering, 1996). Both random and median LHS are available in the Analytica software package (Lumina, 1996).

LHS methods guarantee that values from the entire range of the distribution will be sampled proportional to the probability density of the distribution. Because the distributions are sampled over the entire range of probable values in LHS, the number of samples required to adequately represent a distribution is less for LHS than for random Monte Carlo sampling. LHS is generally preferred over random Monte Carlo simulation (McKay et al., 1979; Iman and Helton, 1988; Morgan and Henrion, 1990). As noted earlier, restricted pairing techniques are available for the purpose of inducing correlations between variables in LHS (Iman and Conover, 1982; Iman and Shortencarier, 1984).

LHS is often referred to as a "variance reduction technique." This is because, compared to Monte Carlo simulation, LHS reduces the statistical fluctuation in simulations of random variables for a given sample size. This is most clearly illustrated by comparing Monte Carlo simulation and LHS for a simple example: the generation of samples from a normally distributed random variable.

Fig. 7.15 illustrates a comparison of median LHS, random LHS, and Monte Carlo simulation for sample sizes of 10, 30, 100, and 1000 applied to the simulation of samples from a normally distributed random variable with mean of 10 and standard deviation of 2.

The differences among the three methods are most clearly revealed for the case of a sample size of 10. In this example, median LHS provides an exact representation of the sampled percentiles of the normal distribution. In this case, 10 equal probability strata were created, and one sample was drawn from the median of each strata. This translates into the following percentiles: 5, 15, 25, 35, 45, 55, 65, 75, 85, and 95. In the case of random LHS, the same 10 strata were created. However, one sample was drawn at random from within each strata. Therefore, there is some random fluctuation in the random LHS case that is not present in the median LHS case. Finally, in Monte Carlo simulation, values are drawn at random and exhibit substantially more statistical fluctuation than for either of the LHS cases.

As the sample size increases, the differences among the three approaches become less pronounced. For sample sizes of 30, 100, and 1000, median and random LHS appear to provide very similar results. However, random Monte Carlo exhibits noticeable random fluctuation for sample sizes of 30 and 100. Even for a sample size of 1000, there is some noticeable random fluctuation in the Monte Carlo simulation compared to the two LHS cases. You can use numerical experiments such as these to explore the behavior of these simulation methods. In this case, the Analytica program was used. Other packages, such as Crystal Ball, allow you to employ either random Monte Carlo simulation or median LHS.

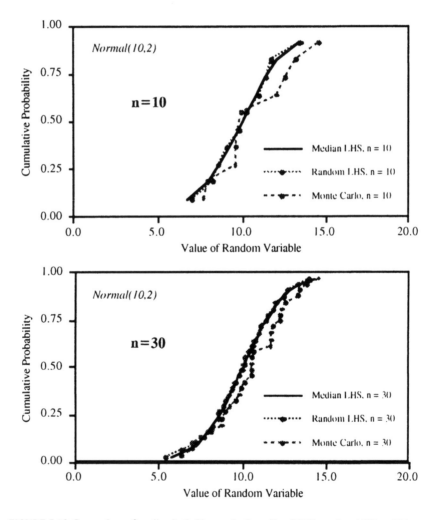

FIGURE 7.15 Comparison of median Latin Hypercube Sampling (LHS), random LHS, and Monte Carlo simulation for 10, 30, 100, and 1000 Samples of a normal distribution with mean of 10 and standard deviation of 2.

   The implications of these numerical experiments are that LHS allows you
to obtain a more accurate representation of the cumulative distribution function
of a specified probability distribution with fewer samples than does random
Monte Carlo. Thus, in many cases, LHS is preferred as a numerical simulation
method. There are some rare pathological cases in which LHS would "break
down" and give erroneous results. For example, if you wished to simulated the
mean of a sine wave and made an unfortunate choice regarding the sample size,
distribution, and domain for the random input and regarding the period of the

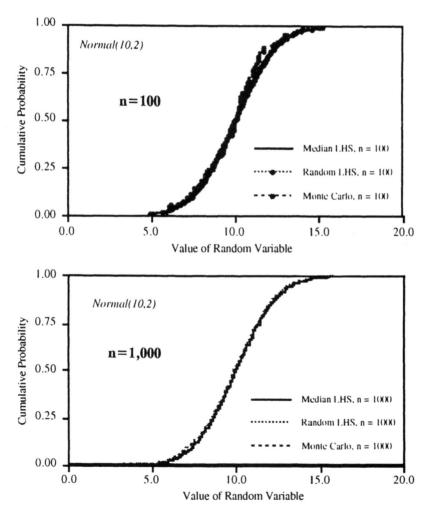

FIGURE 7.15 (*Continued*)

sine wave, you could obtain means of +1 or −1 instead of zero. In most exposure assessment problems, this pathological case is not of concern.

However, there are some perhaps more practical reasons for considering the use of Monte Carlo over LHS. One is if you are interested in characterizing the effects of statistical sampling error. For example, if you wish to use bootstrap simulation to estimate confidence intervals or sampling distributions for the central moments of a data set or the parameters of a parametric distribution (e.g., Efron and Tibshirani, 1993), then you should always use Monte Carlo simulation and never use LHS. We will consider bootstrap simulation more in the discussion of two-dimensional simulations of variability and uncertainty. Also, the methods available for estimating what sample size to use for a numerical simulation are derived based upon classical statistical theory, which is typically predicated upon independent random sampling. We mention some of these methods in Chap. 5. In a strict sense, these methods are only appropriate for Monte Carlo simulation, and not for the LHS methods. However, because LHS typically gives better performance than Monte Carlo simulation, the use of such methods should provide an upper bound on the required number of samples needed from LHS methods. Stated differently, for many purposes LHS methods should perform better for the same sample size compared to Monte Carlo.

There are cases where Monte Carlo methods have been used instead of LHS even for small sample sizes. For example, Hanna et al. (1998) have performed an uncertainty analysis of the Urban Airshed Model (UAM) using Monte Carlo simulation with only 50 iterations of the model. The small sample size was due to the computational intensity of the exercise; the 50 iterations required several weeks of CPU time. In addition, the use of Monte Carlo instead of LHS facilitated the use of many standard statistical tests. For example, using both analytical and numerical approaches, the confidence intervals on the predicted CDFs for peak ozone concentrations were found to be comparable in width to the estimated precision of the model. Thus, the use of a larger sample size would have been unnecessary. This case study illustrates that it is not always possible, appropriate, or necessary to use large sample sizes when propagating distributions through a model.

Three other remarks regarding LHS are needed. One is that the restricted pairing technique developed by Iman and Conover (1982) was created to correct a deficiency observed in LHS. Because LHS involves stratified sampling, it is only random in the pairing of values of the input random variables. It was noticed that large spurious correlations were obtained even when the pairing was done randomly. To reduce these spurious correlations, restricted pairing was introduced and the desired correlations were specified as zero. However, as noted earlier, one would expect to have spurious correlations occur for independent random variables for small sample sizes. A second comment is that LHS methods have limitations with regard to the estimation of system moments. This is a motivation for the use of random LHS over median LHS, or for the use of Monte Carlo simulation instead of LHS methods. While median LHS guarantees that values in the tails of the distribution will be sampled, random LHS allows for the possibility of extreme values to be simulated even within a given strata. This has significance for estimation of the higher central moments, which are sensitive to extreme values. Thus, if you are particularly interested in obtaining good estimates of high percentiles of a distribution, or of the variance, skewness, kurtosis, or other statistic based upon the higher order central moments, you should consider not only large sample sizes, but selection of an appropriate sampling method. Third, LHS simulations may take longer to run in some packages than random Monte Carlo, particularly because of the use of the restricted pairing method to avoid spurious correlations.

7.5.3. Other Sampling Techniques

Random Monte Carlo and LHS are the most common techniques that exposure assessors will encounter. However, other sampling techniques exist. An example that may be of interest to exposure assessors is importance sampling.

Like LHS, importance sampling is referred to as a sample variance reduction technique, because it can provide a more precise representation of a portion of a distribution and many of its statistics than can random Monte Carlo. However, the purpose of importance sampling is not to improve the representation of an entire probability distribution. Rather, it is to improve the estimation of the portion of the distribution with which you may be the most concerned. For example, if you are interested in obtaining a good estimate of the tails of the distribution, then you may wish to generate proportionally more sample points in the tails. This can be done by arbitrarily increasing the likelihood of sampling values at the portions of the input distributions to the model that lead to the tail of the model output. This method requires making transformations based upon the importance weight given to the input samples (Morgan and Henrion, 1990).

7.5.4. Selecting a Sample Size for Numerical Simulations

The sample size corresponds to the number of repetitions used in the numerical simulation. The selection of sample size is usually constrained at the upper end by the limitations of computer software, hardware, and time, and at the lower end by the acceptable confidence interval, or precision, for model results. In cases where the analyst is most interested in the central tendency of distributions for output variables, the sample size can often be relatively small. However, in cases where the analyst is interested in low probability outcomes at the tails of output variable distributions, large sample sizes may be needed using traditional Monte Carlo techniques. As sample size is increased, computer runtime, memory use, and disk use may become excessive. Therefore, it may be important to use no more samples than are actually needed for a particular application. This problem is alleviated in part by new generations of computer hardware, which are characterized by faster processors, more RAM, and larger hard disks than previous generations.

There are a number of approaches that can be used to select the sample size. One approach to selecting sample size is to decide on an acceptable confidence interval for the mean, based upon a preliminary estimate of the standard deviation of the model output. Such an estimate may be obtained, for example, by running a relatively small number of Monte Carlo simulations, calculating the mean and standard deviation at that point, estimating how many additional simulations would be needed to narrow the confidence interval for the mean to an acceptable level, assuming no change in the standard deviation, and then proceeding with the additional simulations.

Another approach to selecting the sample size is based upon developing a confidence interval for whatever fractile level is of most concern in the investigation (Ang and Tang, 1984; Morgan and Henrion, 1990). This analysis can be done for any distributional shape. For example, we may wish to obtain a given confidence that the value of the pth fractile will be bounded by the ith and k^{th} fractiles. In a Monte Carlo simulation, we can use the following relations to estimate the required sample size:

$$i = \text{mp} - c\sqrt{\text{mp}(1 - p)} \qquad (7.21)$$

$$k = \text{mp} + c\sqrt{\text{mp}(1 - p)} \qquad (7.22)$$

The relations in Eqs. (7.21) and (7.22) yield a confidence interval for the pth fractile if the sample size is known, where c is the standard deviation of the standard normal distribution associated with the confidence level of interest. To calculate the number of samples required, the expressions above can be rearranged to calculate the confidence interval $(Y_{p-\Delta p}, Y_{p+\Delta p})$ as follows:

$$m = p(1 - p)\left(\frac{c}{\Delta p}\right)^2 \qquad (7.23)$$

For example, if we wish to be 95% confident that the value of the 95th percentile will be enclosed by the values of the 94th and 96th fractiles, then c would be 2.0, p would be .95, Δp would be 0.01, and m could be calculated to be 1900.

However, another factor to consider in selecting sample size is whether a high degree of simulation accuracy is really needed. In screening studies based on a first-pass set of expert judgments, it may be unnecessary to obtain a high degree of confidence in specific fractiles of the output distribution, because initial estimates of uncertainty may be subject to considerable empirical uncertainty themselves.

The approach to selecting sample size described above is appropriate for use with the Monte Carlo simulation technique. The approach to estimating the precision of modeling results based on confidence intervals will typically overestimate the required sample size needed with LHS.

7.5.5. Verification of Monte Carlo Results

The results of a numerical simulation of probability distributions may be compared with other analyses to identify any potential errors. Alternative methods that may be used include the properties of the mean and variance, central limit theorem, analytical solutions, and approximation techniques described earlier. In addition, interval methods may be used. Interval methods involve placing bounds on each input variable, and then calculating the resulting bounds on the model output. In interval calculations, no other assumptions are made regarding the input variables. For example, it is not necessary to specify a correlation structure, because all possible correlations among the inputs are accounted for by the interval calculations.

An interval variable is bounded by a minimum and maximum as follows (Burmaster and Thompson, 1995):

$$\overline{V} = [\text{min, max}] \qquad (7.24)$$

The bar over the variable denotes this is an interval variable, to distinguish it from other types of variables. If we wish to calculate an exposure using a typical simplified exposure equation of the following form:

$$\overline{E} = \frac{\Sigma_{i=1}^{n} \overline{X}_i}{\Sigma_{j=1}^{m} \overline{Y}_i} \qquad (7.25)$$

Then to calculate the maximum value of exposure, we must use the maximum values for all of the inputs in the numerator and the minimum values of all of the inputs in the denominator. To calculate the minimum value of exposure, we must use the minimum values of all of the terms in the numerator and the maximum values of all of the terms in the denominator.

To illustrate the use of the interval variables, let's consider a simple case study. Suppose that we have an exposure equation of the form of Eq. (7.25), with two terms in the numerator and one term in the denominator. Furthermore, suppose that each term has the following distribution:

$$X_1 \sim N(6,1); \; X_2 \sim N(24,4); \; Y_1 \sim N(14,2) \tag{7.26}$$

The 95% probability of the result, based upon 10,000 simulations using random LHS, is from 6.2 to 15.8.

If we use the 95% probability ranges for each input as the basis for specifying intervals for each variable, we obtain the following assumptions:

$$X_1 = [4, 8], X_2 = [16, 32], \text{ and } Y_1 = [8, 16]$$

and we obtain the following result:

$$\overline{E} = \frac{[4, 8][16, 32]}{[8, 16]} = \frac{[64, 256]}{[8, 16]} = [4, 32] \tag{7.27}$$

This interval encloses the 95% probability range obtained from the probabilistic simulation. Thus, in this case, the interval analysis indicates that the probabilistic analysis is yielding reasonable results.

7.5.6. Available Software Tools

Any summary of software tools is at best a snapshot subject to change over time. However, in spite of the risk that our summary may be obsolete by the time this is read, we venture to mention several packages of either practical or historical interest.

Two of the currently most popular commercially-available packages are @Risk and Crystal Ball. The former is an add-on to the Lotus 1-2-3 or Excel spreadsheet software available for Microsoft Windows system, as well as for Lotus 1-2-3 in DOS and Excel on the Macintosh system (Palisades, 1997), while the latter is an add-on to the Microsoft Excel spreadsheet software available for the IBM PC and Macintosh families of computers (Decisioneering, 1996). Analytica (formerly Demos) is a Macintosh-based graphical environment for creating, analyzing, and communicating probabilistic models for risk and policy analysis (Lumina, 1996). It is now also available for the PC.

Iman and Shortencarier (1984) have developed FORTRAN-based programs for generating samples using Monte Carlo or LHS and for analyzing modeling results using various linear regression techniques (Iman et al., 1985). These programs can be adopted to any modeling platform for which FORTRAN is available.

Several software tools developed by EPA have Monte Carlo simulation capabilities. Of the EPA-supported tools, the most prominent is MOUSE, which is an acronym for Modular Oriented Uncertainty SystEm (Klee, 1992). MOUSE can be used to simulate uncertainties in models consisting of one or more equations. MOUSE has a number of built-in capabilities, including probability distributions, graphics, sensitivity analysis, and so on, to facilitate the development and interpretation of Monte Carlo simulations. The manual for MOUSE also has a good introduction that motivates the need for uncertainty analysis and illustrates some of the key insights obtained.

Ken Bogen has developed a probabilistic modeling system known as RiskQ (Bogen, 1992). RiskQ is implemented in Mathematica. While it has many powerful capabilities, it requires knowledge of programming in Mathematica (Murray and Burmaster, 1993).

The increasing accessibility of user-friendly probabilistic simulation tools has made the probabilistic technique available to larger numbers of people. The graphical interfaces can have something of a hypnotic effect on the new user. It is important, however, to always keep one's modeling and data quality objectives in mind, and to determine based upon those what the appropriate simulation approach is for your specific problem.

7.6. TWO-DIMENSIONAL SIMULATIONS

Historically, Monte Carlo techniques have been used with complex models in the field of nuclear safety and environmental fate. In such analyses, the focus has been on estimating the probability that a system would fail. Furthermore, there have been efforts to estimate the uncertainty regarding the probability of system failure. This has led to a two-dimensional approach to dealing with probabilistic concepts.

In the context of human exposure and risk assessment, two-dimensional probabilistic concepts are gaining increasing attention. The National Academy of Sciences (NRC, 1994) has recommended that the distinction between variability and uncertainty should be maintained rigorously at the level of individual components of a risk assessment (e.g., emissions characterization, exposure assessment) as well as at the level of an integrated risk assessment. A workshop sponsored by the Environmental Protection Agency provided recommendations regarding the use of two-dimensional simulations (USEPA, 1996b), which were incorporated into a 1997 agency policy document (USEPA, 1997a).

7.6.1. Background on Two-Dimensional Methods

Classical theory in statistics has long recognized that uncertainty may exist regarding distributions that describe variability when available data are a random sample from a population (Hahn and Meeker, 1991). For example, sampling distributions and confidence intervals for summary statistics are expressions of uncertainty regarding frequency distributions. Summary statistics may include the moments (e.g., mean, variance) of a data set, parameters of a distribution, percentiles of an empirical cumulative probability distribution, sample correlations, and so on.

As the power of desktop computers increases, numer·~al simulation techniques have been employed more frequently to solve problems regarding confidence intervals. In particular, bootstrap methods, introduced by Efron in 1979, can be employed to characterize sampling distributions and confidence intervals in cases for which analytical solutions may be unavailable (Efron and Tibshirani, 1993). Thus, statistical techniques have existed for some time that address what some analysts refer to as "second-order random variables."

Kaplan and Garrick (1981) present a definition of risk in terms of a triplet including scenarios, probabilities, and consequences. The scenarios may be viewed as variability, while the probabilities may be viewed as uncertainty.

Bogen and Spear (1987) present a number of probabilistic descriptors of risk to different individuals in the population. These include a randomly selected individual, a mean risk individual, a 95th percentile individual, and a "maximum risk" individual. Each individual represents a different sample from the variability in the population distribution. For each of these defined individuals, an estimate of uncertainty in risk is given. A key assumption in the analysis was that the uncertainties faced by all individuals are correlated with each other, such that there is no ambiguity regarding the ranking of their risk compared to other individuals. The validity of this assumption depends, of course, on the specific problem. In addition, other types of dependencies may exist among variable and uncertain quantities and were not addressed. For example, variability in exposures may be independent from one individual to another, and there may be dependencies between uncertain parameters of frequency distributions.

An interpretation of two distinct sources of variation in terms of variability and uncertainty has been made in the nuclear industry and in human health risk assessment. For example, the IAEA (1989) defines "Type A" and "Type B" uncertainties as being due to stochastic variability and lack of knowledge, respectively, and provides an example analysis.

Frey (1992) presented a simulation-based approach to evaluating the distinctions and interactions between variability and uncertainty that relaxes some of the assumptions of Bogen and Spear (1987). The method is based upon a two-dimensional probabilistic simulation, which may be based on sampling techniques such as Monte Carlo, median LHS, random LHS, or others. This approach is able to deal with the case in which uncertainties may be independent from one individual to another. Furthermore, this approach can deal with cases in which the parameters of probability distributions for uncertainties in measurement errors may be conditional on variable quantities.

Hoffman and Hammonds (1994) demonstrated the use of two-dimensional probabilistic simulation for an exposure and risk assessment case study. A method for selecting the sample size of the uncertainty analysis based upon the use of tolerance intervals is given. MacIntosh et al. (1994) employed two-dimensional probabilistic analysis for an ecological risk assessment. Helton (1993; 1996) and others have employed two-dimensional analysis of uncertainty and variability in assessments of radioactive waste disposal methods. Two-dimensional simulations of variability and uncertainty in exposure and risk assessment have also been performed by Cohen et al. (1996), Brattin et al. (1996), and Price et al., (1996a&b).

7.6.2. Distinctions between Variability and Uncertainty in Simulations

In traditional approaches to probabilistic analysis, distributions for both variable and uncertain inputs are propagated through an exposure model in the same dimension of uncertainty. The model output thus represents a hybrid distribution which contains some combination of true variability and uncertainty reflecting a lack of knowledge. Such a distribution can lead to erroneous inferences if the result is misinterpreted.

To illustrate this point, consider three cases illustrated in Fig.7.16. In the first case, suppose that variability dominates over uncertainty. Then the distribution obtained for exposure will reflect the differences in exposure to different members of the exposed population. Hence, the result in this case is a description of variability. In the second case, suppose that uncertainty dominates over variability. The output distribution here represents uncertainty in predicting or estimating the exposure to the population, and the range of uncertainty dominates over the range of variation in exposures to different members of the population. Therefore, the result for the second case is a description of uncertainty, or lack of knowledge. In the third case, both variability and uncertainty contribute significantly to the range of results. We can think of the model output as being two-dimensional in this case, with the result being an uncertain estimate of the frequency distribution for variability in exposures to different members of the population. Thus, for any percentile of the population, we are uncertain about what the actual exposure is. Or, conversely, for any selected exposure level, we are uncertain about what fraction of the population is at or below that level.

7.6.3. Characterization of Uncertainty as an Aid to Model Validation

A practical example of the need to distinguish between variability and uncertainty comes from an international study involving assessment of the accuracy of model predictions for a situation in which both model inputs and the "answer" are known. In a test exercise by the International Atomic Energy Agency, international teams of modelers were asked to make predictions regard-

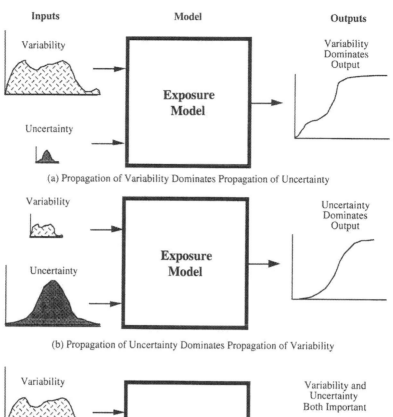

(a) Propagation of Variability Dominates Propagation of Uncertainty

(b) Propagation of Uncertainty Dominates Propagation of Variability

(c) Interactions of Variability and Uncertainty Both Affect Result

FIGURE 7.16 Interactions of variability and uncertainty in an exposure model.

ing interindividual variability of the concentration of ^{137}Cs in humans (Hoffman and Thiessen, 1996). This example is also discussed in Chap. 3. The modelers were given a detailed description of a site during the time of arrival of the Chernobyl plume and information regarding the amount of radionuclides measured in air, soil, and water in 1986. The modelers were asked, among other things, to predict the statistical distribution describing variability among individual whole body concentrations and to also provide a quantitative estimate of uncertainty in terms of 95% subjective confidence intervals for the model predictions. Thus, the modelers were asked to make predictions regarding a true but unknown frequency distribution. The modelers' predictions were then compared to a data set of measured whole body concentrations which were withheld from the modelers by the IAEA to ensure a "blind" test for the model predictions.

The results of the IAEA model test indicated that typically the subjective confidence intervals placed upon the predicted frequency distribution for inter-individual variability in whole body concentrations were too narrow, although in some cases the frequency distribution for interindividual variability was too wide. The net effect of this is that there was in some cases poor agreement between the model predictions and the distribution of measured concentrations. The confidence limits provided by the modelers frequently did not encompass the observations. Thus, the modelers were typically overconfident regarding their results. The explicit treatment of interindividual variability and subjective uncertainty in this study enabled the validity of the model results to be evaluated. However, a limitation of the modeling work in the IAEA study is that there was confusion on the part of some modelers regarding the use of probability distributions to express the state of knowledge about fixed but imperfectly known values versus the use of frequency distributions to represent interindividual variability in whole body concentrations (Hoffman and Thiessen, 1996).

7.6.4. Taxonomy of Variability and Uncertainty in Model Inputs

To expand on the exploratory work by Frey (1992), a general taxonomy is illustrated here to show the methodology by which uncertainty and variability may be incorporated into a single modeling framework. This taxonomy is intended to be from the perspective of an analyst, and it serves to highlight some of the methodological issues associated with quantification of variability and uncertainty. Thompson and Graham (1996) present an alternative view of a similar but less detailed taxonomy from the perspective of a risk manager. This serves to highlight how the results of probabilistic analysis may be used in decision making. The taxonomy from the viewpoint of the analyst includes the following cases:

(1) *Case A: Uncertainty Only.* This is a typical approach to probabilistic analysis, in which all random variables are treated as if they are uncertain quantities, without regard to distinctions between variability and uncertainty.

(2) *Case B: Variability Only.* This case rarely exists in practice. Here, it is assumed that all frequency distributions for variability are known exactly, and that there are no uncertainties. Some probabilistic analyses

are developed and interpreted this way. For example, analyses of exposure or risk that purport to identify high-end exposures may inappropriately comingle uncertainty and variability in a single dimension of probabilistic simulations.

(3) *Case C: Variability with Uncertain Parameters.* This is a common case that unfortunately is not often properly addressed for three major reasons. First, this approach requires a two-dimensional simulation framework. Second, it requires recognition that for small sample sizes, the parameters of a distribution are uncertain. Thus, sampling distributions for the parameters should be characterized, and used as a basis for generating a family of frequency distributions to represent variability. Few practitioners, and even those who recognize the existence of uncertainty in the parameters, properly employ sampling distributions. Third, it requires proper specification of the statistical dependencies between the sampling distributions. This is further described below.

(4) *Case D: Second Order Uncertainty.* This case has been identified in work by Shlyakhter and Kammen (1992) with reference to the treatment of bias and "surprise" in uncertain quantities. This situation may also be used to reflect the lack of precision in estimates of uncertainty based on expert judgment or on small sample sizes (such as in the case of some types of measurement errors based upon replicates).

(5) *Case E: Measurement Error and Variability.* If an imperfect instrument is used to make measurements of a variable quantity, then the set of measurements will include a combination of variation due to the underlying frequency distribution for variability and to the random and systematic error introduced by the measurement itself. Thus, the set of measurements will tend to have a wider range of values, and perhaps a shifted central value, compared to the true distribution for variability. In cases where the measurement error is known, such as when the precision and accuracy of a measurement technique have been quantified, it is possible to separate these sources of variation from the observed values, and to propagate them separately through a model.

(6) *Case F: Generalized Case for Interactions between Variability and Uncertainty.* In general, a set of measurements may be based upon a limited sample, and the precision and accuracy of the measurement technique may be poorly quantified. Thus, there is uncertainty regarding the bias and variance of the measurement and regarding the parameters of the underlying frequency distribution. The generalized case can be complicated by correlation structures among the uncertainties in the parameters for each of the frequency distribution for variability and the probability distribution for uncertainty. Furthermore, the mean and variance of the probability distribution for uncertainty may be a function of the

sample values from the frequency distribution for variability. For example, some measurement errors are proportional to the value of the quantity being measured. In addition, the variability may be correlated with other random variables in a model.

(7) *Case G: Non-Representative Data or Surrogate Data.* In the previous cases listed here, techniques exist for analyzing data to develop quantitative estimates of variability and uncertainty. In this case, uncertainty exists regarding the appropriateness of data with respect to the needs of a particular assessment. For example, available data may not be for a statistically valid random sample. Furthermore, available data may not be for the appropriate averaging times or averaging areas. In the case of a study of chronic health or ecological effects, distributions of long term averages for intake rates, pollutant concentrations, and other model inputs are needed, whereas for a study of acute health or ecological effects, distributions of short term averages are needed. Yet available data may be for other averaging times. In addition, available data may only be a surrogate for that which is actually needed for a specific study. In cases such as these, statistical analysis may be inadequate for fully capturing uncertainties regarding variability. Thus, expert judgment may be needed to quantify uncertainties in these situations.

7.6.5. General Approach for Simulation of Variability and Uncertainty

A two-dimensional approach to Monte Carlo simulation can be employed to properly disaggregate and evaluate the consequences of variability and uncertainty, as well as to simulate their interactions such as for Cases C, E, and F as described in the previous section. Such an approach is only necessary when both variability and uncertainty are important. The two-dimensional approach is also referred to by some as "nesting" or "double looping." This approach is gaining attention and is seen in an increasing number of applications in exposure assessment (Frey, 1992; Hoffman and Hammonds, 1994; Carrington, 1993; Cohen et al., 1996; Frey and Rhodes, 1996, and others as previously noted). In a simulation approach, it is possible to relax some of the assumptions that are required for an analytical approach to be tractable. For example, various types of dependencies between variable and uncertain inputs can be handled numerically that may be difficult or impossible to handle analytically. Therefore, the discussion that follows is based upon the use of simulation.

An exposure model may be written as (borrowing terminology from Bogen and Spear, 1987):

$$E = E(V, U) \tag{7.28}$$

The estimate of exposure to the population of exposed individuals is a function of variability in parameters V, which have different values for each member of the population, and uncertainty in parameters U for which there is some lack of information about their true values. In general, it is possible for multiple uncertain quantities to be correlated with each other. It is also possible for the uncertainties, U, to be correlated or dependent on the variable quantities, and for variable quantities, V, to be dependent on each other.

Given a general model $E = f(V,U)$ as described above, the first step is to disaggregate the model variables into variable and uncertain components, as illustrated in Fig. 7.17. For all variable quantities, frequency distributions must be specified. For all uncertain quantities, probability distributions must be specified. A sampling technique such as LHS is then employed to generate two sets of samples. For each of the M variable quantities, the frequency distributions are simulated with a sample size of m. For each of the N uncertain quantities, the probability distributions are simulated with a sample size of n. In principle, the sample sizes m and n for the variable and uncertain dimensions of the simulation need not be the same. However, because of the interactions between variability and uncertainty, it is often preferable that these sample sizes be the same. Thus, the sample size for the two-dimensional simulation is $m \cdot n = m^2 = n^2$. Clearly, this can impose a potentially severe computational burden, depending on the required sample size.

The model is repetitively evaluated for each combination of samples from the variable and uncertain parameters. In cases where the uncertainties are independent from one individual to another, the rank ordering of individuals is also uncertain. This important interaction is captured by the two-dimensional approach to simulation, because for every sample of values for the vector of uncertain quantities, a separate frequency distribution for variability is simulated. Therefore, there will be n estimates of the rank, or percentile, representing a probability distribution for the rank of the individual within the population. In contrast, one-dimensional simulation approaches do not capture these interactions. The resulting hybrid frequency/probability distribution for exposure is only meaningful if interpreted to represent an individual selected at random from the population, or if only variability or uncertainty dominates. Otherwise, it is inaccurate to draw conclusions from such results regarding the rank ordering of individuals within the population, or the exposure level faced by an individual at a given fractile of the population.

7.6.6. Developing Input Assumptions for Two-Dimensional Analyses

As noted earlier, in some cases random variables may be unambiguously subject to either variability or uncertainty, but not both. However, in many cases,

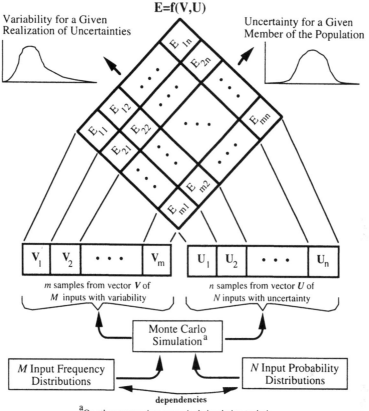

$$E = f(V, U)$$

Variability for a Given
Realization of Uncertainties

Uncertainty for a Given
Member of the Population

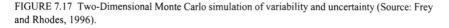

FIGURE 7.17 Two-Dimensional Monte Carlo simulation of variability and uncertainty (Source: Frey and Rhodes, 1996).

a given model input may be both variable and uncertain. Here, we will focus on Case C of the taxonomy given in the previous section. Frey and Rhodes (1996) address Cases B, C, and E. Chap. 9 also includes discussion of Cases C and E.

Case C addresses the situation in which uncertainty exists due to limited sample data. As an example, we consider a case study similar to that reported by Frey and Rhodes (1996) involving prediction of hazardous air pollutant (HAP) emissions for coal-fired power plants. The example presented here is for a different type of power plant boiler and particulate matter control device than the one used by Frey and Rhodes (1996). Specifically, we will focus on estimation of manganese emissions from a front-fired, coal-fueled boiler with a downstream fabric filter for particulate matter control (Frey, 1998b). Data from USEPA

(1996d) are available regarding HAP concentrations, and the partitioning of HAPs in the boiler and particulate matter control devices for 11 key HAPs. Partitioning refers to the mass flow rate of a HAP in the flue gas exiting a device divided by the mass flow rate of the same HAP entering the same device. By knowing the concentration of HAPs in the coal, the coal flow rate into the boiler, and the partitioning of the HAP in the boiler and downstream devices, it is possible to estimate the HAP emissions at the stack. A key limitation, however, is that the measurements required to estimate partitioning of each type of HAP in each type of device (e.g,. boiler, particulate matter collection) are expensive. Therefore, few data are available to support modeling efforts. The data that are available represent averages over a three-day period. The model that is used in this example is the same as that used in Frey and Rhodes (1996).

There are only three data points available for the partitioning of manganese in front-fired boilers. It is assumed that these data are representative of the quantity that we truly wish to model, and that the data are a reasonably random sample from the overall population. In addition, it is assumed for the time being that measurement error is negligible compared to other sources of uncertainty. Thus, it is assumed that the most significant source of uncertainty is random sampling error because we have only three data points.

In Fig. 7.18, the three data points are shown plotted as an empirical distribution. A dashed line is shown to illustrate the beta distribution that was fitted to the three data points. A beta distribution was selected because the partitioning factor is a quantity that must be bounded by zero and one. The graph indicates that the distribution provides a good fit to the data. However, because there are only three data points, there is uncertainty regarding the true value of the parameters of the beta distribution and regarding the mean and variance that we

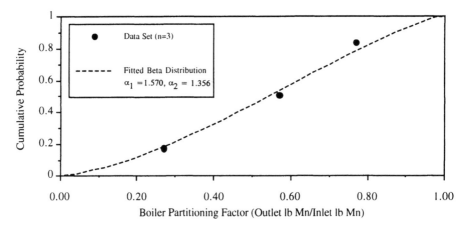

FIGURE 7.18 Comparison of an empirical distribution of data with a fitted beta distribution.

would obtain if we collected more data. Therefore, there is uncertainty regarding what the shape of the beta distribution would be if we collected more data.

To evaluate uncertainties due to sampling error, a number of methods may be used. In Chap, 5, we discussed some methods that can be used in special cases to estimate confidence intervals for the mean of many types of distributions and data sets and for the variance of a normal distribution. Here, we are interested in estimating the uncertainty in the parameters, α_1 and α_2, of the beta distribution, and of the mean and variance of the beta distribution. In cases for which analytical solutions may not exist, it is possible to develop sampling distributions and confidence intervals for statistics using bootstrap simulation (Efron and Tibshirani, 1993).

To illustrate how bootstrap simulation may be used to quantify uncertainty in the parameters of a distribution, we describe its application to the partitioning factor data. We employ what we refer to as a "parametric bootstrap," in which we assume that the fitted distribution is the best estimate of the true distribution, and we seek to characterize the uncertainty in the parameters of the fitted distribution. To do this, we numerically simulated "replications" in which we draw a random sample of three data points from the fitted distribution. For each replication of three data points, the mean, variance, α_1, and α_2 were calculated. Monte Carlo simulation was used for this purpose, because we seek to mimic the random fluctuations in data values that we would obtain if we experimentally measured a random sample of three front-fired systems. The number of replications to be simulated was chosen to be 500. This was a compromise between the gain in accuracy that would be obtained from additional replications versus computational time and analysis effort. Efron and Tibshirani (1993) note that for the purpose of estimating confidence intervals for some statistics (e.g., mean), 200 replications may be sufficient. However, larger numbers of replications may be needed in other cases (e.g., for a confidence interval on an extreme percentile).

The fitted distribution was simulated 500 times, using a FORTRAN code originally developed by Rhodes (1997), to generate 500 bootstrap replications of three independent data points each. The 500 replicates of these moments and parameters were then plotted as scatter plots and CDFs. Fig. 7.19 displays the scatter plot of α_2 versus α_1. This plot is a way to represent the bivariate distribution for these two parameters. The plot illustrates two main points. One is that both α_1 and α_2 are subject to considerable uncertainty when only three data points are available. α_1 ranges from approximately 0.01 to nearly 1000, while α_2 has a similar range. The second main point is that there is a dependence between α_1 and α_2 that would be difficult to capture using other methods. There is a positive linear correlation between the parameters. The linear correlation between the two is 0.79. However, conditional variance of α_2 is not constant with respect to different values of α_1.

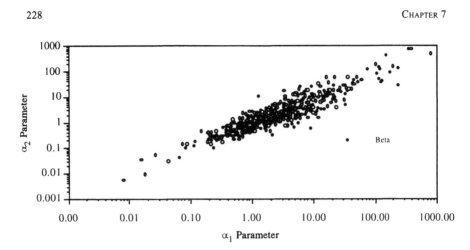

FIGURE 7.19 Scatter plot of the uncertainty in α_1 and α_2 for a beta distribution fitted to three data points.

The marginal sampling distributions for the α_1 and α_2 parameters are shown in Figs. 7.20 and 7.21, respectively. In each case, the marginal sampling distributions are compared to the best estimate obtained from the method of moments. These graphs demonstrate that the best estimate is only one plausible estimate regarding the true values of the parameters, and that many other values are possible.

A similar set of analyses were performed for the mean and variance of the fitted Beta distribution. A scatter plot of the mean and variance is shown in

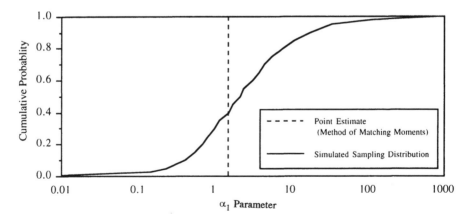

FIGURE 7.20 Comparison of the marginal distribution obtained from bootstrap simulation of α_1 of a beta distribution versus the best estimate of α_1 obtained from the method of moments.

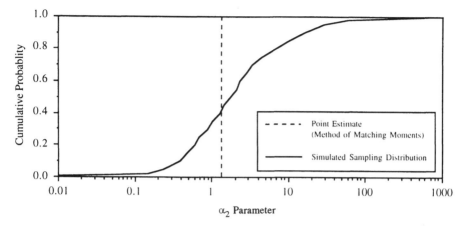

FIGURE 7.21 Comparison of the marginal distribution obtained from bootstrap simulation of α_2 of a beta distribution versus the best estimate of α_1 obtained from the method of moments.

Fig. 7.22. It is clear from the scatter plot that there is a dependence between the variance and the mean. Because the beta distribution has finite lower and upper bounds, there are constraints on the possible set of joint values of the mean and variance. This is reflected in the upside-down "u" shape to the maximum values obtained for the variance as a function of different values of the mean. In this case, the linear correlation coefficient is –0.20. However, the dependence is not linear, nor is it even monotonic. This type of nonlinear and nonmonotonic dependence between the variance and the mean is revealed by bootstrap simulation. The marginal sampling distributions for the mean and variance are com-

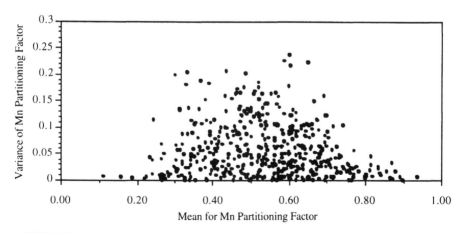

FIGURE 7.22 Scatter plot of the uncertainty in the mean and variance of beta distribution fitted to three data points.

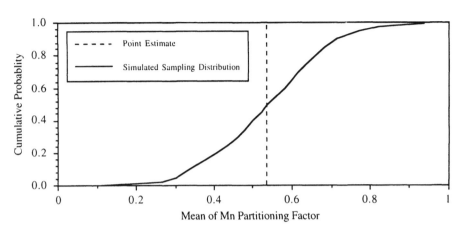

FIGURE 7.23 Comparison of the marginal distribution obtained from bootstrap simulation of the mean of a beta distribution versus the best estimate of the mean.

pared to the best estimates in Figs. 7.23 and 7.24, respectively. In both cases, it is apparent that there is significant uncertainty in the mean and variance. In fact, the uncertainty in the mean appears to cover approximately the same range of values for a 95% probability range as does the original data set.

The implications of the uncertainties in the parameters of the beta distribution for the partitioning factor were evaluated in a probabilistic simulation of variability and uncertainty. In the first part of the simulation, the parameters α_1 and α_2 were estimated for each of the 500 bootstrap samples using the method of matching moments. Thus, a total of 500 parametric distributions were fitted

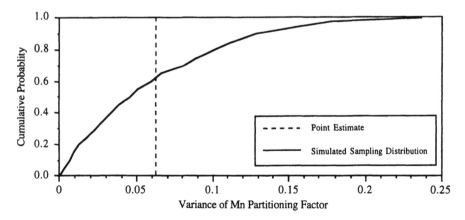

FIGURE 7.24 Comparison of the marginal distribution obtained from bootstrap simulation of the mean of a beta distribution versus the best estimate of the mean.

to each of the 500 bootstrap samples. For each of the 500 parametric distributions, a total of 500 samples were simulated in a second loop of the probabilistic simulation, in order to characterize the CDF of each alternative distribution. Thus, a total of 250,000 samples were simulated (500 samples for each of 500 fitted distributions). Then, the results of the two-dimensional simulation were analyzed to construct probability bands representing uncertainty in the fit of a beta distribution to the three original data points.

The result of the two-dimensional simulation of the partitioning factor is shown in Fig. 7.25. The dark grey areas represent the 50% probability band of the results, and the light grey areas depict the 99% probability band. The extreme upper and lower bands represent the 99% probability band. The original three data points are shown as large dots. With a data set comprised of only three data points, there is clearly a substantial amount of uncertainty in the estimate of the partitioning factor, as represented by the wide uncertainty bands for the fitted distribution.

The front-fired boiler partitioning factor for manganese was one of several inputs to a model of power plant manganese emissions. Other inputs include manganese concentration in the coal, partitioning of manganese in the fabric filter, which were both variable and uncertain, and coal heating value, plant capacity factor, and the power plant heat rate. The latter three were assumed to represent variability with no uncertainty, because a substantial amount of data were avail-

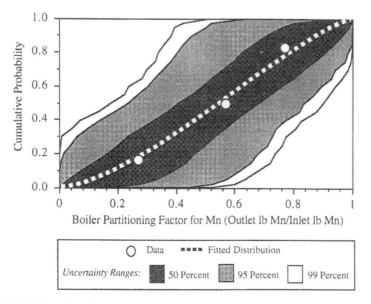

FIGURE 7.25 Two-dimensional representation of variability in Mn boiler partitioning factor and uncertainty due to limited sample size.

able for these, and measurement errors were negligible. The details of the assumptions for the latter three inputs are reported in Frey and Rhodes (1996). In Table 7.3, a summary is given of input assumptions for not only manganese, but a total of 11 HAPs, which were considered in a series of detailed case studies (Frey, 1998b). Here we present a few results to illustrate the types of findings that may be obtained from two-dimensional simulation.

The two-dimensional result for the emissions of manganese for a three-day averaging period is shown in Fig. 7.26. These results depict 500 realizations of the frequency distribution for the variability of manganese emissions from one three-day averaging period to another. The results indicate that there is significant uncertainty regarding the true frequency distribution for variability in emissions.

For any given value of emissions, it is possible to read from the graph the 90% probability band regarding the fraction of three-day periods that would have emissions less than or equal to a specified value. For example, there is a 90% probability that between 80 and 99% of the three-day averaging periods have emissions less than 50 lbs. Similarly, there is a 90% probability that the emissions of the 95th fractile of the three-day averaging periods will be between 10 and 170 lbs.

Fig. 7.27 shows the variability and uncertainty in emissions of the sum of 11 HAPs, based upon the input assumptions in Table 7.3. There is a 95% probability that between approximately 88 and over 99% of the three-day period will have emissions less than or equal to 1000 lbs. The 95th percentile of three-day periods have a 95% uncertainty range in emissions from approximately 200 lbs to well over 1500 lbs.

Fig. 7.28 displays results of more direct policy interest, and also provides a comparison of emissions estimates based upon the use of alternative assumptions for probability distributions fitted to coal concentration data. Fig. 7.28(a) displays the uncertainty in annual average emissions of the sum of 11 HAPs. Each annual average is calculated based upon an entire frequency distribution for variability in emission from one three-day period to another. The emission rate was also converted from units of lbs per year to tons per year. Because one year contains approximately 122 three-day periods, single average was calculated based upon 122 randomly sampled values of three-day emissions from each of the 500 frequency distributions simulated in the two-dimensional method. Since 500 three-day periods were sampled from each of the 500 distributions, it was possible to develop four sets of annual averages (based upon $4 \times 122 = 488$ samples from each of the 500 frequency distributions). Thus, Fig. 7.28(a) shows four separate distributions representing uncertainty in annual average emissions. Each distribution represents uncertainty. However, the differences from one distribution to another reflect variability in the annual average from one year to another.

An emission limit of 25 tons per year for the sum of 189 HAPs has been considered for application to electric power plants. Fig. 7.28(a) indicates that, if

TABLE 7.3 Summary of Input Assumptions for Assessment of Variability and Uncertainty in Hazardous Air Pollutant Emissions (Source: Frey, 1998b)

Description	1st Parameter[a]	2nd Parameter[a]	Sample Size (n)[b]
Power Plant Characteristics			
Heat Rate	9780	580	Normal Distribution
(BTU/kWh)			No Uncertainty
Coal-Heating Value	10000	180	Normal Distribution
(BTU/lb)			No Uncertainty
Capacity Factor	N/A	N/A	Mixture Distribution
Coal Concentrations (ppmw), Fitted by Gamma Distributions			
An - Antimony	0.177	14.589	12
As - Arsenic	1.526	7.585	19
Be - Beryllium	2.862	0.530	18
Cd - Cadmium	0.393	32.969	7
Cr - Chromium	5.813	2.617	21
Co - Cobalt	2.608	2.083	15
Hg - Mercury	5.297	0.0224	17
Mn - Manganese	1.877	15.236	19
Ni - Nickel	2.911	5.997	19
Pb - Lead	1.270	8.155	17
Se - Selenium	0.444	7.211	20
Boiler Partitioning Factors (lb out/lb in) Fitted by Beta Distributions			
An - Antimony	0.550	1.346	3
As - Arsenic	0.128	0.052	3
Be - Beryllium	1.184	0.628	3
Cd - Cadmium	0.432	0.081	3
Cr - Chromium	1.071	0.160	3
Co - Cobalt	140.16	4.335	3
Hg - Mercury	0.683	0.216	3
Mn - Manganese	1.570	1.356	3
Ni - Nickel	5.352	0.969	3
Pb - Lead	0.229	0.198	3
Se - Selenium	0.249	0.128	3
Fabric Filter Partitioning Factors (lb out/lb in) Fitted by Beta Distributions			
An - Antimony	1.663	48.215	3
As - Arsenic	1.203	83.706	6
Be - Beryllium	0.366	23.114	5
Cd - Cadmium	0.184	0.145	4
Cr - Chromium	0.260	3.897	5
Co - Cobalt	1.294	247.61	5
Hg - Mercury	0.430	0.171	5
Mn - Manganese	0.440	10.560	6
Ni - Nickel	0.255	5.532	6
Pb - Lead	0.803	82.684	6
Se - Selenium	0.194	0.465	4

[a]For normal distribution, parameters are mean and standard deviation. For gamma distribution, parameters are shape and scale. For beta distribution, parameters are α_1 and α_2.
[b]Sample size, where reported, is used as a basis for bootstrap simulation to estimate uncertainty.

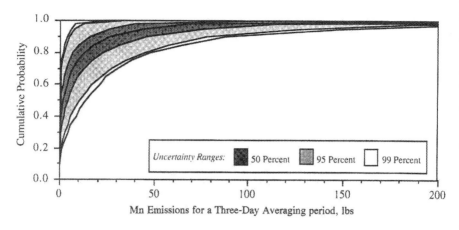

FIGURE 7.26 Variability and uncertainty in manganese emissions.

only 11 of those HAPs are considered, there is approximately a 95 percent probability of compliance with such a standard for bituminous coal-fueled front-fired boilers equipped with fabric filters. However, the results presented in Fig. 7.28(a) are based upon the use of gamma distributions to represent variability in coal concentrations, as documented in Table 7.3. Instead, if lognormal distributions are used, then the results given in Fig. 7.28(b) are obtained. Since lognormal distributions tend to be more "tail-heavy" than the gamma distribution, the results indicate a higher probability of exceeding the possible standard. The variability in annual averages from one year to another appears to be sufficiently small that the estimates of uncertainty are robust to variability.

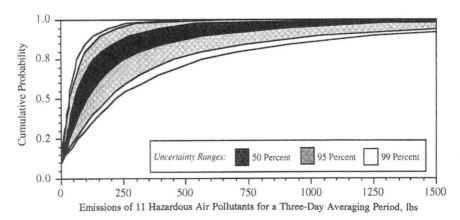

FIGURE 7.27 Variability and uncertainty in the sum of emissions for 11 hazardous air pollutants.

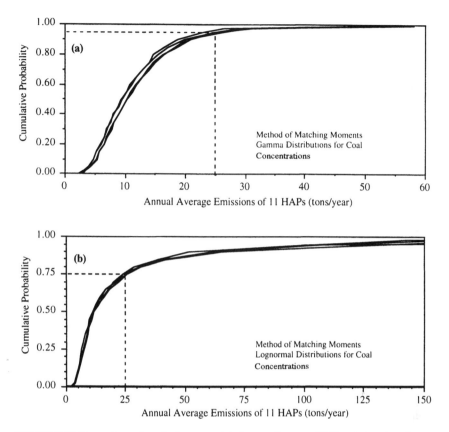

FIGURE 7.28 Comparison of results for uncertainty in annual average HAP emissions based upon different parametric distributions used to represent variability and uncertainty in coal pollutant concentrations: (a) Gamma distributions fitted to coal concentration data; (b) Lognormal distributions fitted to coal concentration data.

The similar case study reported by Frey and Rhodes (1996) addresses other issues regarding the interpretation of the two-dimensional simulation results. These include: sensitivity and statistical analysis to identify the key input assumptions that most significantly contribute to variability and uncertainty in the model outputs; characterization of uncertainty in annual emissions based upon the simulation of three-day averages; and comparison to one-dimensional simulations.

Other factors to consider in quantitative analysis of variability and uncertainty include the effect of selection of parametric distributions and alternative parameter estimation methods. For example, Frey and Rhodes (1998) compare and evaluate variability and uncertainty based upon alternative parametric distributions (e.g., lognormal, gamma, and Weibull) commonly used to represent non-negative quantities, such as concentrations, that are often part of exposure asseessments. Frey and Burmaster (1998) show how the choice of parameter estimation methods, such as method of matching moments versus maximum likelihood estimation, can lead to different estimates of uncertainty for a given parametric distribution.

7.6.7. Simulating Correlations among Frequency Distributions for Variability in Two-Dimensional Analyses

In the previous section, one of the topics addressed was correlations among uncertainties in the parameters of frequency distributions for variability. In this section, we illustrate how to address correlation among different frequency distributions in the context of a two-dimensional simulation. This example is based upon the bivariate normal distribution using techniques described in Section 7.5.1.4. The example also draws upon the use of bootstrap simulation to characterize uncertainty in the parameters of a frequency distribution.

Consider a simple model of the form $Z = X + Y$. Both X and Y are normally distributed random variables, with a mean of 10 and a standard deviation of 2. However, only a finite number of data points, $n = 10$, are available from which the mean and standard deviation were estimated. Therefore, there is uncertainty regarding the true mean and variance for each distribution. Furthermore, there may be correlation between these two variables. In a series of case studies, we evaluate the implications of different correlations between X and Y on the sum Z.

For the purpose of characterizing uncertainty in the mean and variance of the normal distributions, several methods can be used. In this case, there are analytical solutions available for the uncertainty in the mean and variance in the form of Student's t and chi-square distributions, respectively. For illustrative purposes, however, we choose to use the numerical method of bootstrap simulation to estimate uncertainty in these parameters. As noted earlier, bootstrap simulation can be used in cases for which analytical solutions may not be available. Using bootstrap simulation, joint values of the mean and variance were created for both X and Y. Then, for each joint value of mean and variance, a frequency distribution was simulated. The correlation between each realization of frequency distributions for X and Y was simulated using the bivariate normal distribution. Four cases of correlations were considered. Three involved specification of point estimates for the correlation. These three values were −0.9, 0, and +0.9. In addition, a case was considered in which the correlation was itself assumed to be uncertain, and was specified to vary uniformly from the minimum to the maximum possible values (−1,1). The results for Z are shown in Fig. 7.29.

As expected, in the case of the negative correlation of −0.9, the narrowest range of variability in Z was obtained. In this case, the range of uncertainty is comparable to the range of variability. As the correlation coefficient increases,

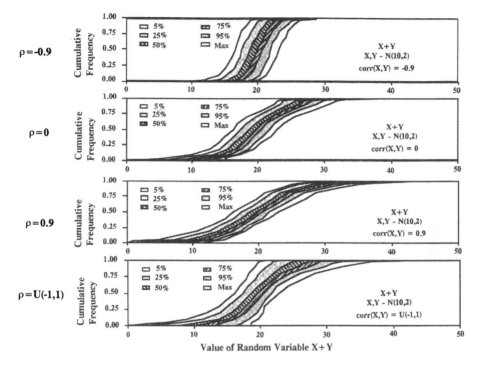

FIGURE 7.29 Simulation of the sum of two normal distributions (Mean =10, Standard Deviation = 2) based upon a sample size of 10 and 200 replications and four cases of correlations between variable quantities.

the range of variability also increases. The range of uncertainty for any given percentile of the frequency distribution actually does not change significantly; the 90% probability band has a range of approximately 10 units in these three cases. However, when the correlation coefficient is assumed to be uncertain, the range of uncertainty in the tails of the result for Z becomes much wider than for the other three cases. Furthermore, the case of uncertain correlation appears to capture all of the possible outcomes that are contained in the three previous cases. Thus, an advantage of two-dimensional simulation of variability and uncertainty is that it can also enable simulation of the effect of uncertainty in correlations among frequency distributions.

7.7. DISCUSSION OF ANALYTICAL, APPROXIMATION, AND NUMERICAL METHODS

We have reviewed several methods that can be used to estimate uncertainty and/or variability in model outputs based upon specification of information regarding these sources of variation in model inputs. The methods include implications of the central limit theorem, properties of the means and variance for multiplicative and additive models, analytical solutions for model output distributions based upon the method of transformation of variables, approximation methods based upon Taylor series expansions, and numerical methods. Numerical methods we have focused upon include Monte Carlo simulation and LHS. Other methods, such as importance sampling, were briefly mentioned. We also considered how numerical methods may be used for two-dimensional analyses of both variability and uncertainty. We have not exhaustively covered all possible methods, but we have presented some of the most common ones.

Of the methods described, perhaps the most appealing is the transformation of variables method, because it provides an exact solution for the distribution of model outputs. However, its most significant drawback is that it can be applied to only simple problems and, therefore, is typically not of practical interest. Of the other methods, numerical simulation comes closest to this ideal. The major advantages of numerical simulation are that it can accommodate a wide variety of assumptions regarding model inputs and can be used with a wide variety of models. Numerical methods are also available for evaluating the effect of correlations and dependencies among model inputs, as we have described. The accuracy of numerical methods can be improved by increasing the simulation sample size.

The other methods reviewed provide only limited information regarding the distribution for a model output, because they provide estimates of the central moments and not of the entire distribution. This is true, for example, of the Taylor series expansion methods. These methods require the calculation of partial derivatives of the model, and can often be done numerically. However, unless a sufficient number of terms are included in the expansions, these methods can give erroneous results even for relatively simple nonlinear functions, as we have illustrated here.

A potentially significant feature of the exact and approximation methods reviewed to this point is that it can be relatively straightforward to identify the key sources of random variation in a model output. For example, consider an additive function which is the sum of three independent random variables, X_1, X_2, and X_3. Suppose that the variances of the three variables are 50, 30, and 20, respectively. It is then possible to calculate, using Eq. (7.2), that the variance of the sum is 100. Furthermore, one can infer that 50% of the variance in the sum is attributable to X_1, 30% of the variance of the sum is attributable to X_2, and only 20% of the variance in the sum is attributable to X_3. This is powerful information which can be used to prioritize additional data collection or research, in combination with consideration of the costs of research and the potential for specific research programs to lead to uncertainty reductions. For example, it would seem intuitive that it may be worthwhile to attempt to reduce uncertainty in the model input that most contributes to uncertainty in the model output, which in this case is X_1. Similarly, it may be worthwhile to better characterize uncertainty in X_1 by collecting more data, and not to spend additional time or effort refining the estimate of uncertainty for X_3. Of course, it is also useful to consider the value of information for each of the model inputs. It may be the case that it is either very expensive or not possible to reduce uncertainty in a particular input.

As the models used deviate from the linear function of independent random variables, it becomes more difficult to develop a unique measure of the relative contribution of each model input to the variance of the model output. In Chap. 8, we discuss a number of methods that can be used in conjunction with numerical simulation techniques to identify the key inputs that could be targeted for additional data collection or research.

The methods based upon propagation of moments share a significant shortcoming: knowledge of only the mean, variance, and other moments of a model output is often not of central importance to decision-makers. In exposure and risk assessment, it is typically important to understand the behavior of the tails of the probability distribution for a model output. The approximation methods based upon Taylor series expansion can be extended to make estimates of higher order central moments of the model outputs. Knowledge of a sufficient number of the central moments of the model output can be used to select a distribution, such as from the Pearson or Johnson families, for the model output. Such a distribution can then be used to make approximate estimates of the tails of the model output distribution. These types of analyses require information regarding the higher order central moments of the model inputs.

In many practical cases, the higher order central moments can be extremely difficult to estimate with precision. For example, it is well known that estimates of the mean and variance of normal and lognormal distributions can be highly uncertain when only a small number of data points are available. However, the third or fourth central moments will typically be even more difficult to estimate. Various approximation methods exist for calculating confidence intervals regarding the skewness and kurtosis (Bowman and Shenton, 1986). For example, the uncertainty in the skewness of a sample of data drawn from a normal distribution may be sufficiently large that it can be difficult to determine whether the data have no skewness, or may in fact be from a positively or negatively skewed distribution. Similarly, the estimates of the kurtosis may be sufficiently uncertain that it can be difficult to select an appropriate shape for the distribution without considering other relevant information regarding the data (e.g., that it is a physical quantity that must be non-negative, that it is the result of a particular biological or mathematical process, etc.).

Approximation methods are based upon a significant number of assumptions regarding truncation of the Taylor series and the central moments of the model inputs. Numerical simulation methods have an analogous but different set of comparable assumptions. The length of the Taylor series used for any given system moment translates into computational effort proportional to the number of random variables. In contrast, the computational effort of numerical simulation methods, such as simple Monte Carlo, is not dependent on the number of input variables. Instead, it can be specified based upon the precision with which an analyst requires a prediction of a statistic of the model output. The statistic may be the mean, variance, a given percentile, or other. The inability to precisely specify the distributions for inputs to a model is a challenge for simulation methods, just as it is for approximation methods. These challenges can be addressed in a number of ways. For simulation methods, one approach, previously discussed, is a two-dimensional method for distinguishing between variability and uncertainty.

7.8. ASSESSING MODEL UNCERTAINTIES

In the previous sections of this chapter, we have focused on methods for propagating variability and/or uncertainty through a model. In this section, we briefly discuss an approach to evaluating uncertainties in model structure. As noted earlier, uncertainties in model structure may be evaluated by comparing alternative models using techniques such as decision trees or logic trees. Evans et al. (1994a) employed a probability tree to consider various types of model and structural uncertainties in a formaldehyde risk assessment. For example, the top level of the probability tree involves a judgment as to whether formaldehyde should be classified as a carcinogen. The second level involves the weighting of alternative mechanistic hypotheses regarding chemical carcinogenesis. Examples of the mechanistic hypotheses include chronic cell proliferation, direct genotoxicity, or a combination of the two. Another level involves alternative adjustments for interspecies extrapolations. The probability weights assigned to each alternative model at each level of the tree may be dependent on previous levels.

Concerns in dealing with model uncertainties include trying to include an exhaustive set of alternative model formulations in the assessment as well as the development of weighting factors in a probability tree or similar approach. It may be confusing, for example, as to what a probability weight placed upon an alternative model really means. Does it refer to a probability of a model being correct? As noted earlier, while it may be possible to falsify a model, it is difficult or impossible, just as it is with any scientific hypothesis, to state that a model is "true." However, a model may be found to be reasonably adequate for a given purpose. Thus, a probability weight might be interpreted as the likelihood with which a model, after further study, will be found to be reasonably and distinctly adequate compared to other models.

7.9. UNCERTAINTY REDUCTION TECHNIQUES

Approaches for reducing uncertainty typically involve collecting more data or information that enables development of better data sets or models. However, the nature of an assessment can also affect the level of uncertainty. For example, analyses of short-term exposures related to acute health effects must address uncertainty in predicting concentrations, intake rates, and exposed populations over many short time intervals. For example, ambient concentrations of carbon monoxide near a roadway will exhibit substantial variability on an hourly basis due to changes in traffic flow, traffic speeds, vehicle mix, wind fields, ambient temperature, and other factors. However, daily average carbon monoxide concentrations may tend to be relatively similar for all weekdays. The analysis of long-term exposures related to chronic effects would not need to consider the wide range of possible values that would occur in any specific short time period, but rather would be focused upon sources of variation that would affect the long-term average values. In the case of the carbon monoxide example, uncertainty in long term average values would be associated with limited sample sizes of ambient monitoring data, and daily or seasonal variability that may lead to uncertainty in the sampling distribution for the annual average. Analyses that involve averaging or aggregation will typically be confronted with lower levels of uncertainty than those analyses which must consider small populations or short averaging times.

Similarly, aggregation of populations can reduce the amount of uncertainty that must be considered in an analysis. For example, the estimate of the transfer rate of a chemical from a pasture to a cow, as described in Chap. 4, would depend on whether the analysis was being done for a randomly selected cow at a randomly selected pasture or for an average herd of cows and an average pasture. In the former case, the random combination of a cow and a pasture would be represented by the entire frequency distribution for all cows and all pastures, whereas in the latter case we would only have to consider the sampling distribution of the mean transfer rate, and not the full range of interindividual variability for each cow. The range of uncertainty for an average cow would typically be much less than the range of uncertainty for a randomly selected cow.

CHAPTER 8

IDENTIFYING KEY CONTRIBUTORS TO VARIABILITY AND UNCERTAINTY IN MODEL OUTPUTS

There are multiple motivations for identifying model inputs which are key contributors to variability and uncertainty in model outputs. First, an identification of significant contributors to output variance gives the analyst an awareness of which inputs "drive" the results. Further, a basic exploration of the models, inputs and results, promotes improved understanding and interpretation of the analysis. Finally, this exercise is a first step in the prioritization of future research and information gathering.

In most probabilistic assessments, the majority of the variance in the output distribution is attributable to variability and/or uncertainty in a small subset of the inputs (Belcher and Travis, 1991; Frey and Rubin, 1992a; McKone and Bogen, 1992; Taylor, 1992; Cullen, 1995; Frey and Agarwal, 1996). An identification of this subset of highly significant contributors to output variance has useful implications for future research. It enables us to target resources for the characterization of uncertainty in a small number of important inputs, rather than spread resources thinly across the entire set of inputs entering the analysis. For linear models, the output distribution (particularly its upper percentiles) tends to be dictated by the input parameters with the largest coefficients of variation. Inputs with high degrees of skewness or nonlinear relationships to the output are often most influential in nonlinear models. Beyond examining coefficients of variation, there are a variety of techniques for identifying key contributors including scatter plots, correlation coefficients (including partial and rank correlations), multivariate regression, contribution to variance, and probabilistic sensitivity analysis, each of which is outlined briefly below. Additional information is available in the following sources (Bevington and Robinson, 1992; Iman and Conover, 1982; Draper and Smith, 1981; Iman and Helton, 1988; Ishigami *et al.*, 1989; Morgan and Henrion, 1990; Iman and Hora, 1990; Helton, 1993;

Kalagnanam *et al.*, 1998). Additional information about the error propagation properties of various types of models is discussed in Section 3.1.

In the examples discussed below we assess the importance of inputs on the entire range of the output distribution in one-dimensional analysis. Also, we discuss a method for gauging the relative importance of variability and uncertainty as well as specific sources of uncertainty (e.g., random sampling error) in two-dimensional analysis. In more sophisticated analyses it may be desirable to assess the importance of inputs on a specified portion of the output distribution, e.g., the upper 95th percentile. Those interested in this area are directed to the NCRP Commentary No. 14 (NCRP, 1996) where an example is presented. For a discussion of additional issues that arise in two-dimensional analysis, such as uncertainty in correlation coefficients for key sources of variability and variability in correlation coefficients for key sources of uncertainty, see Frey (1992); and Frey and Rhodes (1996).

After identifying model inputs which contribute significantly to variability and uncertainty in the output it is necessary to consider what to do with this information. We recommend value of information (VOI) techniques to assess the importance of the variability and uncertainty contributed by individual inputs in the decision at hand. Basic VOI theory, an extension of uncertainty analysis and application of Bayes' Theorem, was developed in the late 1950s (Raiffa and Schlaifer, 1961; Raiffa, 1968; March and Simon, 1958; DeGroot, 1969). In recent years, many individuals have applied the principles of VOI analysis to environmental decision-making (Henrion, 1982; Evans, 1985; Finkel and Evans, 1987; Evans *et al.*, 1988; Reichard and Evans, 1989; Lave *et al.*, 1988). VOI techniques seek to identify situations in which the cost of reducing uncertainty is outweighed by the benefit of the reduction. In the prioritization of research targeted at reducing uncertainties in individual inputs, this means considering the cost and effectiveness of pursuing the reduction of these uncertainties, given the particulars of the available decision alternatives. In short, VOI is helpful in identifying model inputs that are significant in two ways: (i) they contribute significantly to variance in output; and (ii) they change the relative desirability of the available alternatives in the decision under consideration.

8.1. INTRODUCTION TO THE EXAMPLES

In this chapter, specific techniques are illustrated using two case studies. Both of the case studies involve the estimation of average daily dose of semi-volatile contaminants to human receptors via inhalation of airborne contaminants and ingestion of contaminants in locally grown produce. The first case is concerned with estimating the average daily dose of PCBs to an adult female living within three miles of a contaminated marine harbor in New Bedford, Massachusetts, during an approximately one-year period of sediment dredging. The second case concerns the estimation of the long term average daily dose of contaminant to an adult living and working in close proximity to a municipal waste incinerator, hypothetically placed in Connecticut (Cullen, 1995).

8.1.1. Case A

The first case study (Case A) is presented in detail in Chap. 9. Measurements of the concentration of PCBs in air and locally grown produce serve as the basis for exposure assessment. Two models are used to combine concentration data and intake rates (Eqs. (9.3) and (9.4)). Information about the distributions representing these model inputs in the analysis appears in Table 9.2 and Chap. 9.

8.1.2. Case B

The second case study (Case B) is an assessment of daily dose of a contaminant emitted by a municipal waste incinerator, to an individual living in the vicinity, averaged over the life of the facility. In this case, a contaminant emanating from a proposed facility is under assessment, and thus concentrations in air and locally grown produce are estimated from models, rather than measurements as in Case A. We present Case B only to illustrate techniques for the identification of significant contributors to variability and uncertainty. Two intake models are presented here in a consolidated form (Eqs. (8.1) and (8.2)). The first model is used to estimate intake of a chemical contaminant via inhalation. The dilution factor of the contaminant is modeled using a Gaussian plume representation. The maximum receptor concentration is used in this analysis. Average daily dose is estimated assuming that the receptor breathes only outdoor ambient air. The second model is used to estimate intake of a chemical contaminant via ingestion of locally grown root produce, such as, potatoes. The concentration of contaminant in locally grown produce is represented by a simple compartmental model.

In this section we include some basic information about the structure of the models and the distributions used to represent the inputs. The parameters of the selected distributions are presented using the terms introduced in Chap. 4. In the original work the airborne load of contaminant is divided into subportions denoted, j, to represent particles of various diameters, and a vapor phase, for the development of distributions for deposition velocity, rate of retention in the lung and emission rate (Cullen, 1995). We refer interested readers to this source for more complete information.

Inhalation Intake (*Ih*)

$$Ih = \sum_j IrDr_j DfQ_j \tag{8.1}$$

where,

> *Ih* = daily intake via inhalation (mg/day),
> *Ir* = inhalation rate, (m³/day), lognormal ($m = 20$, $\sigma_{\ln(x)} = 0.18$),
> Dr_j = deposition efficiency of particle bound or gaseous contaminant in the lung (–), uniform [bounds vary by category j],
> *Df* = dilution factor, (concentration in ambient air, C_{air})/(incinerator emission rate, *Q*), (mg/m³ per g/s), lognormal ($m = 0.00017$, $\sigma_{\ln(x)} = 0.34$), and
> Q_j = contaminant emission rate from incinerator (g/s), total emission rate is distributed loguniform[$a = 1.5 \times 10^{-6}$, $b = 5 \times 10^{-6}$], with separate distributions for each category, j.

Ingestion Intake of Locally Grown Root Produce (*Ig*)

$$Ig = \left(\sum_j Q_j Df Vd_j \right)\left(M\left[\frac{(1 - \exp(-Wt))If}{WY} + \frac{U}{bD\rho} \right] \right) \tag{8.2}$$

where,

> *Ig* = Intake via ingestion of locally grown root produce (mg/day),
> Vd_j = deposition velocity of airborne contaminant (m/day), loguniform [bounds vary for each category, j]
> *M* = daily consumption of locally grown root produce, (g/day), lognormal($m = 12.5$, $\sigma_{\ln(x)} = 0.6$),
> *W* = weathering constant (1/day), loguniform[$a = 0.01$, $b = 0.1$],
> *t* = time to harvest (days), uniform[$a = 40$, $b = 60$],
> *If* = interception fraction for plant surfaces (–), uniform[$a = 0.05$, $b = 0.25$],
> *Y* = yield of root produce (kg/m²), uniform[$a = 6$, $b = 10$],
> *U* = uptake factor for incorporation of contaminant in soil into root tissue (–), uniform[$a = 0.25$, $b = 1.0$],
> *b* = decay constant of contaminant in soil (1/day), uniform[$a = 0.0001$, $b = 0.0002$],
> *D* = mixing depth (m), uniform[$a = 0.15$, $b = 0.25$], and
> ρ = soil density (g/cm³), lognormal[$m = 1.4$, $\sigma_{\ln(x)} = 0.14$].

8.2. METHODS TO USE PRIOR TO SIMULATION

Summary statistics provide many opportunities for gauging which inputs are likely to be important contributors to variance in a simulation. If a model is linear, the magnitude of an input's variance and/or coefficient of variation gives a rough indication of its potential significance and contribution to output variance.

8.2.1. Apportioning Variance by the Gaussian Approximation

The magnitudes of the variances of the inputs in a simulation provide an initial indication of their significance. One very common variance apportionment technique relies on a first order (or Gaussian) approximation. Using this approximation the variance in a model's output is estimated as the sum of the variances of the inputs, weighted by the squares of any constants which may be multiplied by the inputs as they occur in the model (Bevington and Robinson, 1992). This assessment may be carried out prior to (or apart from) simulation in many cases.

For models in the form of products of inputs, the variances of the log-transformed inputs are the relevant parameters. The fraction of the variance in the natural logarithm of the output attributable to each input is easily calculated. If the inputs are lognormal, loguniform, or logtriangular establishing the variance of the log-transformed inputs is quite straightforward. In the event that an input is uniformly or normally distributed, a lognormal approximation is available (Broadbent, 1956). This approach is most useful for simple model structures into which inputs represented by one of these common distributional forms are entered.

For example, inhalation intake in Case B (Eq. (8.1)) is a product of four random variables, of which Ir and Df are distributed lognormally, Dr is distributed uniformly, and Q is distributed loguniformly. Thus, the fraction of the variance attributable to each input can be decomposed as follows:

$$\ln(Ih) = \ln(Ir) + \ln(Dr) + \ln(Df) + \ln(Q) \tag{8.3}$$

and,

$$\sigma^2_{\ln(Ih)} = \sigma^2_{\ln(Ir)} + \sigma^2_{\ln(Dr)} + \sigma^2_{\ln(Df)} + \sigma^2_{\ln(Q)} \tag{8.4}$$

and finally,

$$\frac{\sigma^2_{\ln(Ih)}}{\sigma^2_{\ln(Ih)}} = \frac{\sigma^2_{\ln(Ir)}}{\sigma^2_{\ln(Ih)}} + \frac{\sigma^2_{\ln(Dr)}}{\sigma^2_{\ln(Ih)}} + \frac{\sigma^2_{\ln(Df)}}{\sigma^2_{\ln(Ih)}} + \frac{\sigma^2_{\ln(Q)}}{\sigma^2_{\ln(Ih)}} = 1 \tag{8.5}$$

TABLE 8.1 Summary Statistics for Inhalation Intake Model Inputs (Case B)

Input	Mean μ_x	Standard deviation σ_x	Variance ($\ln(x)$) $\sigma^2_{\ln(x)}$	Coefficient of Variation σ_x/μ_x
Ir	20.34	3.74	0.033	0.184
Dr	0.82	0.043	0.0027	0.053
Df	0.0002	6.2E-5	0.11	0.34
Q	3.0E-6	9.8E-7	0.12	0.33

Some disadvantages of this technique include the potential for significant error as a result of the use of the first order approximation in cases where the inputs and output are not linearly related. Further, this approach is restricted to consideration of a limited range of values of the inputs. Finally, additional terms would be required to handle correlation structures among inputs.

For example, Table 8.1 contains the summary statistics for the inputs to Eq. (8.1). The variances of $\ln(Df)$ and $\ln(Q)$ are significantly larger than those for $\ln(Ir)$ and $\ln(Dr)$, thus there is an initial indication that Df and Q may be important contributors to variance.

8.2.2. Coefficient of Variation

Another approach to the identification of potentially influential inputs involves comparing the coefficients of variation from each input distribution. A coefficient of variation, or relative standard deviation, is a standard deviation that has been scaled by the mean, i.e., σ/μ (see Section 5.1). This measure may help reduce the potential for attributing an unjustified degree of importance to inputs that take relatively large values. Inputs with large coefficients of variation are candidates for the most significant contributors to output uncertainty in a probabilistic simulation for a linear model. The coefficient of variation also is useful for identifying significant uncertain factors *prior* to simulation or to assess how large the uncertainty or variability in an input *would have to be* to influence the output distribution appreciably. When an analyst has little information or experience relevant to a specific input it may be valuable to reflect on the coefficients of variation of other inputs, to get a sense of the importance of this lack of knowledge.

It should be noted that this method must be used with caution. It is useful only for linear models in which all inputs have approximately the same scale with respect to the output. For example, consider:

$$y = a_1 x_1 + a_2 x_2 \tag{8.6}$$

If x_1 and x_2 have the same coefficient of variation, but $a_1 \gg a_2$, then x_1 dominates the overall uncertainty. Even if the coefficient of variation for x_1 is less than for x_2, x_1 may still dominate where $a_1 \gg a_2$.

Returning to Case B, we note that the coefficients of variation for Df and Q are quite large (Table 8.1), thus these two inputs potentially are key contributors to variance in the output distribution (Eq. (8.1)). Use of the more sophisticated approaches discussed below is necessary before definitive statements about importance may be made.

8.3. METHODS TO USE AFTER SIMULATION

8.3.1. Scatter Plot

A simple visual assessment of the significance of the influence of individual inputs on an output is possible using scatter plots. Each realization in a Monte Carlo or other probabilistic simulation generates one value for each input and output. The simulated values can be plotted in two dimensions, each plot posing a selected input against the output. Scatter plots are most revealing when variation in the output is dominated by variation in a small number of inputs. Linear and nonlinear relationships, as well as patterns and trends, may be hypothesized from the scatter plots and then tested using other techniques outlined here. The use of scatter plots to reveal nonlinear and monotonic relationships is discussed in Chap. 5 (Fig. 5.23) and Chap. 7 (Fig. 7.22).

Scatter plots of simulated values of selected pairs of inputs and outputs from the intake models are presented for illustration (Figs 8.1 through 8.4). These are based on one dimensional simulations of variability and uncertainty. Three of the models are linear in form (Eqs. (9.3), (9.4), and (8.1)). The fourth is nonlinear (Eq. (8.2)); however, because it is dominated by a small number of inputs acting linearly, the relationships between inputs and output behave generally linearly (Cullen, 1995).

8.3.1.1. Case A. Scatter plots of simulated values suggest a very strong relationship between inhalation intake (In) and the concentration in indoor air (C_i), but no such relationship with concentration in outdoor air (C_o) (Fig. 8.1). This result may be attributed to the fact that individuals are spending the majority of their time indoors where contamination in air is approximately a factor of 10 greater than outdoors.

A moderate association is observed between the concentration of contaminant in vine produce (C_v) and ingestion intake (Ig) (Fig. 8.2). A weak association is observed between ingestion intake and the consumption rate in vine produce (M_v) (Fig. 8.2). It appears that little or no influence is exerted on the output by the other inputs.

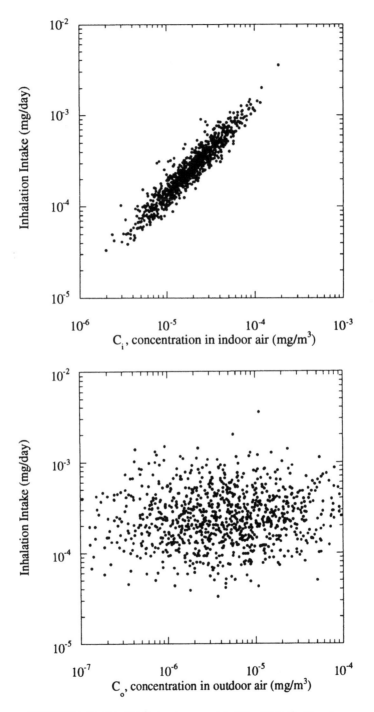

FIGURE 8.1 Scatter plots for inputs versus inhalation intake for Case A.

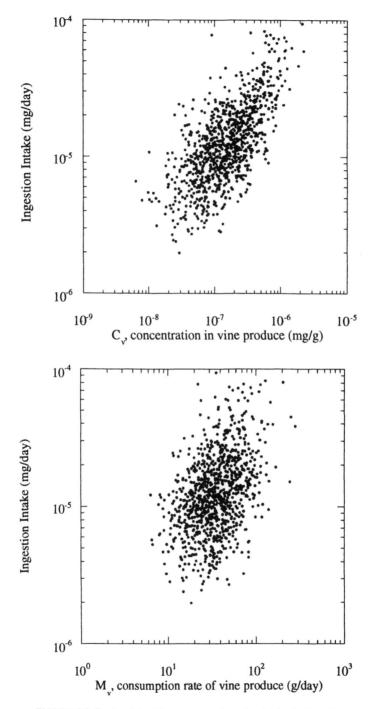

FIGURE 8.2 Scatter plots of inputs versus ingestion intake for Case A.

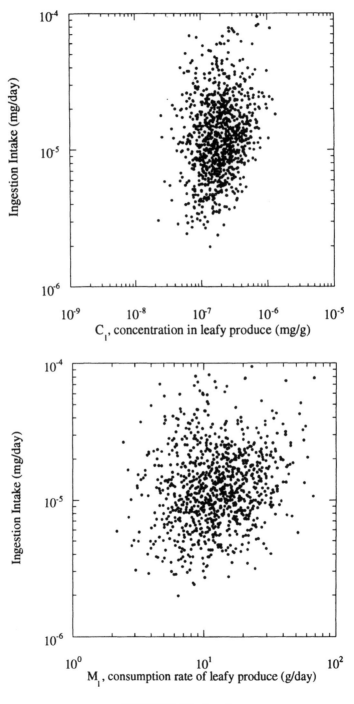

FIGURE 8.2 (*Continued*)

8.3.1.2 Case B. We observe a relatively strong association between inhalation rate (Ir), dilution factor (Df), and emission rate (Q) and inhalation intake (In), while deposition efficiency (Dr) shows little influence (Fig. 8.3).

We observe a strong association between gas deposition velocity (Vd) and ingestion intake (Ig), while the other inputs appear to be weakly related, if at all (Fig. 8.4).

8.3.2. Correlation Coefficient

A correlation coefficient is a measure of the degree of association or covariance between two random variables. For our purposes, correlation coefficients provide an estimate of the linear dependence of a model output on a particular model input. Sample correlation coefficients are sensitive to two factors: (1) the strength of a linear relationship between the input and output; and (2) the range of variation of the output relative to the range of variation of the input. A general discussion of correlation among multiple input factors appears in Chap. 6.

Correlation coefficients are estimated based on the sample values of both the inputs and output, and their respective means (Eq. (8.7)):

$$\rho_{x,y} = \frac{\sum_{k=1}^{m} (x_k - \bar{x})(y_k - \bar{y})}{\left(\sum_{k=1}^{m} (x_k - \bar{x})^2 \sum_{k=1}^{m} (y_k - \bar{y})^2 \right)^{1/2}} \tag{8.7}$$

where m is the sample size (i.e., number of iterations in the simulation), x is an input, y is an output, and x_k and y_k are sample values of x and y. The value of the correlation coefficient, $\rho_{x,y}$, may vary from -1 to 1 (Fig. 8.5), where:

$\rho_{x,y} = 1$ implies linear dependence, positive slope (y increases as x increases),
$\rho_{x,y} = 0$ implies no linear dependence, thus the value of x provides no useful information about the value of y, and
$\rho_{x,y} = -1$ implies linear dependence, negative slope (y decreases as x increases).

For a specific example assume that 1000 iterations of a simulation have been carried out in Case B (Eq. (8.2)). The correlation coefficient between Q and Ig can be calculated as follows:

$$\rho_{Q,Ig} = \frac{\sum_{k=1}^{1000} (Q_k - \bar{Q})(Ig_k - \overline{Ig})}{\left(\sum_{k=1}^{1000} (Q_k - \bar{Q})^2 \sum_{k=1}^{1000} (Ig_k - \overline{Ig})^2 \right)^{1/2}} \tag{8.8}$$

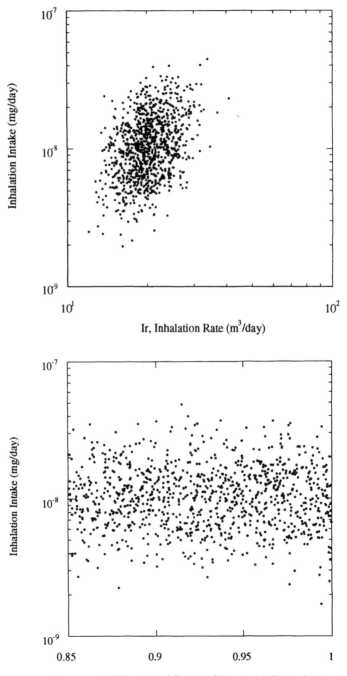

FIGURE 8.3 Scatter plots of inputs versus inhalation intake for Case B.

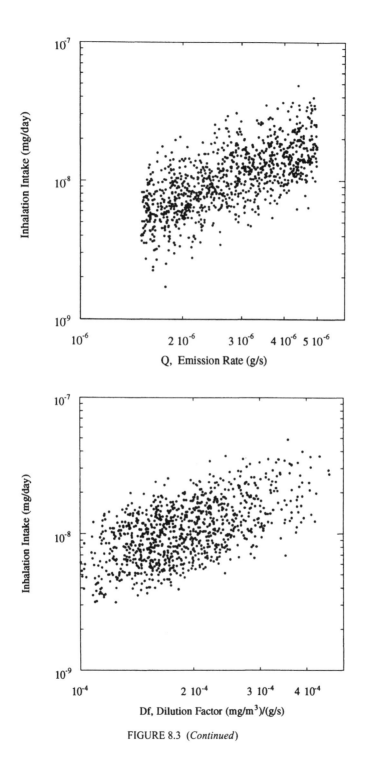

FIGURE 8.3 (Continued)

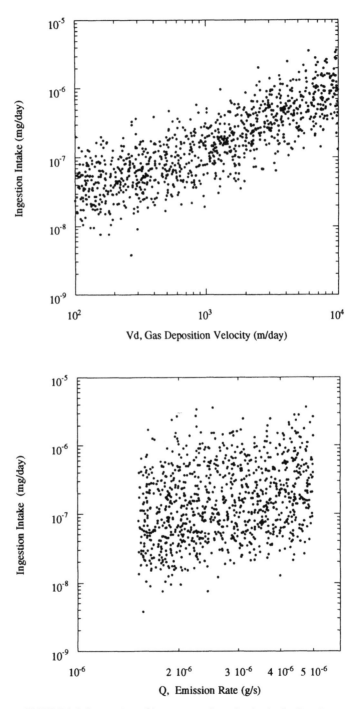

FIGURE 8.4 Scatter plots of inputs versus ingestion intake for Case B.

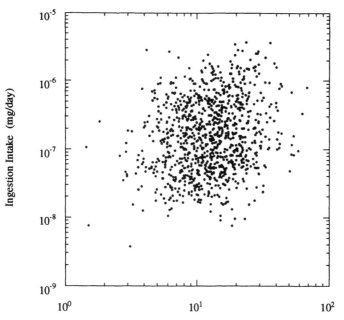

Mr, Consumption Rate of Locally Grown Root Vegetables (g/day)

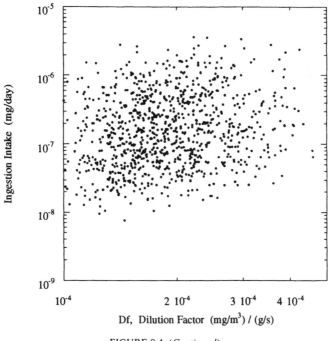

Df, Dilution Factor (mg/m^3) / (g/s)

FIGURE 8.4 (*Continued*)

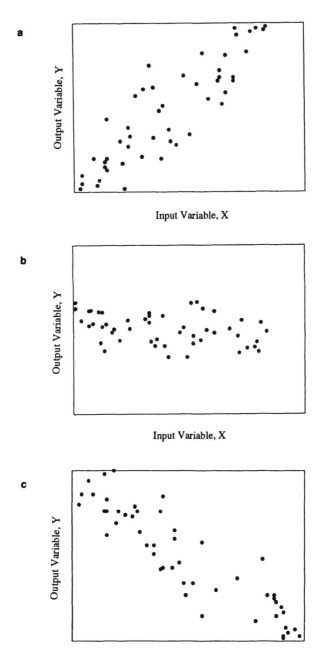

FIGURE 8.5 Examples of correlation between an input and an output. (a. Positive correlation, b. No correlation, c. Negative correlation).

For typical probabilistic simulations, correlations between an input and an output of greater than $|0.5|$ indicate substantial dependence of the variation in the output on variation of the input. Values as low as $|0.2|$ also may be of interest, depending on the sample size. The strength of positive correlations is illustrated in Fig. 8.6. Larger correlation coefficients indicate less dispersion of sample values from an idealized linear relationship between an input and an output. As noted in Chap. 7, the confidence intervals for correlation coefficients can be wide for small simulation sample size (see Fig. 7.12).

The effect of the range of variation of a model input on the magnitude of a correlation coefficient is illustrated in Fig. 8.7. In Fig. 8.7a, the correlation coefficient is calculated based on the full range of data points for the input. However, if the input varies only across a limited range, i.e., between the dashed lines as shown in Fig. 8.7b, then the correlation coefficient will change. This comparison emphasizes that correlation coefficients are sensitive both to the strength of the linear relationship between an input and output as well as to the range of variation of the input.

There are times when the use of a sample correlation coefficient is not appropriate or informative. For example, if several values of the input or output are clustered together in a random pattern, and one point is located away from the cluster, an artificially high estimate of the correlation may result. Alternatively, in cases in which the underlying dependence is nonlinear (but monotonic), such as between skin surface area and body weight, the sample correlation coefficient will underestimate the dependence. One way to avoid these problems is to calculate the correlation coefficient using the ranks of both x and y, as explained below.

Each sample of a random variable (or input) has a rank. The rank is determined by ordering the sample values in ascending order (discussed in Section 5.3.1). The lowest sample value has rank one, the next lowest has rank two, and so on. The largest sample value has a rank equal to the number of samples. If two inputs are related by a monotonic function then the rank ordering of input samples will map onto the rank ordering of output samples in a one-for-one linear relationship. In this case, the ranks of the two inputs will be related by a linear function, resulting in a rank correlation coefficient of 1 (or -1). Thus, the rank correlation coefficient is robust to different monotonic dependencies between an input and output. This is illustrated graphically in Fig. 8.8, where the sample and rank correlation coefficients are developed for the same underlying functional relationship between x and y. Whereas a sample correlation coefficient would indicate only a moderate linear correlation, a rank correlation coefficient would correctly characterize the strong non-linear dependence by being equal to or almost equal to positive unity.

For illustration, sample and rank correlation coefficients for simulated output from Cases A and B were generated and are summarized in Tables 8.2 to 8.5 respectively. These statistics are reported for all of the model inputs with the exception of those in the ingestion model in Case B, where only the four most influential inputs are included (Eq. (8.2)).

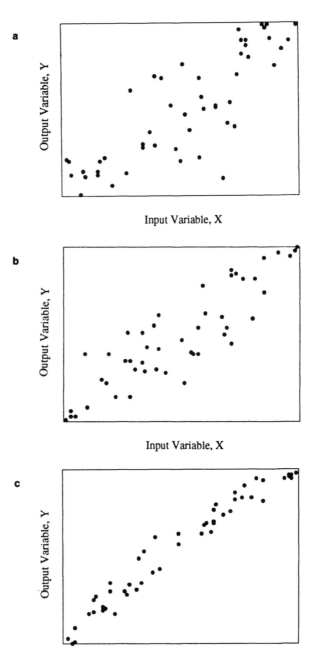

FIGURE 8.6 A qualitative view of correlation. (a. Weak positive correlation, b. Moderate positive correlation, c. Strong positive correlation).

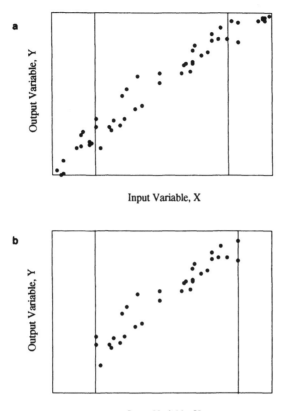

FIGURE 8.7 Effect of range of variation of input on the correlation coefficient. (a. Strong positive correlation, b. Weak positive correlation).

Case A:

The input most strongly correlated with inhalation intake is the concentration in indoor air (C_i) (Table 8.2). This result is consistent with the indication of the scatter plots.

TABLE 8.2 Sample and Rank Correlation Between Inputs and Inhalation Intake (Case A)

Input	Sample Correlation Coefficient	Rank Correlation Coefficient
Ir	0.26	0.27
C_o	0.16	0.12
C_i	0.94	0.95
Fi	0.06	−0.01

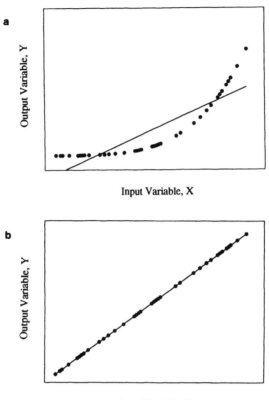

FIGURE 8.8 Sample correlation versus rank correlation. (a. Sample data plot, b. Rank plot).

TABLE 8.3 Sample and Rank Correlation Between Inputs and
Ingestion Intake (Case A)

Input	Sample Correlation Coefficient	Rank Correlation Coefficient
M_r	0.08	0.20
M_v	0.33	0.40
M_l	0.11	0.16
C_r	0.08	0.21
C_v	0.81	0.66
C_l	0.13	0.22

TABLE 8.4 Sample and Rank Correlation Between Inputs and Inhalation Intake (Case B)

Input	Sample Correlation Coefficient	Rank Correlation Coefficient
Ir	0.37	0.38
Dr	0.00	0.00
Df	0.59	0.62
Q	0.69	0.63

The inputs most strongly correlated with ingestion intake are the concentration and consumption rate of vine produce, C_v and M_v respectively (Table 8.3). Sample and rank correlation coefficients yield consistent results for these examples and confirm the relationships suggested by the scatter plots.

Case B:

The inputs most strongly correlated with inhalation intake are the dilution factor (Df) and the emission rate (Q). Sample and rank correlation coefficients produce similar results in this case and are consistent with the results of other techniques.

The input most strongly correlated with ingestion intake is deposition velocity (Vd), with lesser but striking correlations for the other three inputs. Eq. (8.2) has a non-linear form, thus we observe some differences in the strength of sample correlation and rank correlation between the inputs and ingestion intake. For example, the importance of the dilution factor (Df) and the emission rate (Q) is reversed when ranks are used, although the effect is slight and the correlation coefficients for these terms are quite similar.

8.3.3. Correlation Ratio

An alternative to the sample and rank correlation coefficients is the "correlation ratio" a phrase coined by Glaeser et al. (1994). This index is helpful when

TABLE 8.5 Sample and Rank Correlation Between Inputs and Ingestion Intake (Case B)

Input	Sample Correlation Coefficient	Rank Correlation Coefficient
Vd	0.71	0.83
M_r	0.18	0.19
Q	0.21	0.24
Df	0.17	0.29

uncertainty in model inputs is due to uncertain inputs and alternative model formulations, a case in which many other importance measures may be incapable of ordering inputs. The correlation ratio, $r(Y,X)$, is defined:

$$r(Y, X) = \sqrt{\frac{Var(E(Y|X))}{Var(Y)}} \tag{8.9}$$

which is derived from the relationship,

$$Var\{Y\} = E(Var(Y|X)) + Var(E(Y|X)), \tag{8.10}$$

where E stands for expectation and Var stands for variance. Where X and Y are independent, $E(Var(Y|X))=Var(Y)$ and $Var(E(Y|X))$ is zero. In the case of dependence between Y and X, the opposite result holds. Correlation ratios are always positive; therefore they are not useful for establishing the direction of the correlation relationship.

8.3.4. Multivariate Linear Regression

Among the more powerful and flexible techniques for identifying key contributors to variance are those utilizing the principles of multivariate regression. An overview of key concepts is presented here while basic details of multivariate regression can be obtained in other texts (Edwards, 1984; Chatterjee and Price, 1991; Draper and Smith, 1981; Ang and Tang, 1984; Dillon and Goldstein, 1984; Weisberg, 1985; DeGroot, 1986; Iman and Helton, 1988; Morgan and Henrion, 1990).

In general, regression analysis is used to help understand the interrelationships between and within a set of random variables. The application of regression techniques is often pursued in order to gain insights, and hence does not always include the use of extensive formal statistical analysis. In fact in many practical applications, the assumptions underlying the most commonly used forms of regression analysis are not strictly satisfied. Thus, some statistical tests, along with professional judgment and the availability of data, are used to guide the selection of inputs, the representation of relationships in the regression models, and validation of the models. The "goodness" of the regression models is indicated with common summary statistics, graphical comparison of the model predictions with the actual data, and evaluation of the appropriateness of the model relationships with *a priori* expectations.

8.3.4.1. Partial Correlation Coefficients. A partial correlation coefficient (PCC) is a gauge of the strength of the linear relationship between an output and an input. It is estimated using the results of multivariate linear regression analysis. In PCC analysis, the input most highly correlated with the output is assumed as the starting point for construction of a linear model. In the regression model, the output is treated as the dependent variable and the most highly correlated input is treated as a predictive variable. The partial correlation technique then searches for the input which is most highly correlated with the residuals of the regression model already containing the first input. The residual is the difference between the actual sample value of the dependent variable and the estimated sample values, using the linear regression model containing the first input. The process is repeated to add more variables in the analysis. In short, the partial correlation coefficient is a measure of the unique linear relationship between the inputs and outputs that cannot be explained by variables already included in the regression model. This is a computationally intensive approach. It is useful when the relationship between inputs and outputs is monotonic and in the absence of large variations in scale between inputs and output.

The multivariate regression approach involves fitting the parameters of a linear model (Draper and Smith, 1981; Iman and Shortencarier, 1985). For example, for Eq. (8.1) a linear model would take the form:

$$Ih = a_0 Ir + a_1 Dr + a_2 Df + a_3 Q \tag{8.11}$$

where the coefficients, a_i, are measures of the linear sensitivity of inhalation intake to the inputs, Ir, Dr, Df, and Q.

8.3.4.2. Standardized Regression Coefficients. A second regression-based estimate, measuring the importance of the contribution of individual inputs to overall variance in the output, is the standardized regression coefficient (SRC) (Morgan and Henrion, 1990). This measure is recommended for cases in which the scales of the inputs differ widely, as an alternative to the partial correlation coefficient derived from multivariate regression:

$$SRC = \frac{a_i \sigma_{Ig}}{\sigma_i} \tag{8.12}$$

where σ_i and σ_{Ig} are the standard deviations of the inputs, x_i, and the output, Ig, intake via ingestion, respectively. The results are identical regardless of whether the parameters are normalized before analysis, i.e., by subtracting the mean and dividing by the standard deviation, $(x_i - \bar{x})/s_x$. For a linear model in which all of the variance in the output can be explained by the input variables, the sum of the squared standardized regression coefficients is 1, i.e., $\sigma_{Ig}^2 = \sum_i (a_i \sigma_i)^2$. The square of each standardized regression coefficient is interpreted as the fraction of the output uncertainty attributable to the associated input. These techniques are quite simple to use and understand; however, since functional relationships between the output and many of the basic inputs are nonlinear, only a portion of the variance can be explained by the linear regression model. Iman et al. (1985) refer to SRCs as portraying the shared contribution of model inputs, while PCCs identify the unique contribution of an input to variance in the output.

8.3.4.3. Rank Regression Coefficients. A variation on these regression-based measures explores the relationship between the ranks rather than the individual values of model inputs and output. The PRCC (partial rank correlation coefficient) and SRRC (standardized rank regression coefficient), calculated from the ranks of the simulated values rather than the values themselves, are recommended over the PCC and SRC when nonlinear relationships, widely disparate scales, or long tails are present in the inputs and outputs. Ranks are valuable replacements for simulated values in such cases because the limitations of the SRC and PCC preclude their usefulness. In particular, SRC and PCC are based upon least squares regression, which requires that residuals be normally distributed, whereas approaches using ranks are not constrained in this way.

Using simulation results from 1000 iterations for Case B, partial correlation coefficients and standardized regression coefficients were obtained via multiple regression for ingestion intake (Eq. (8.2)). The regression R^2 was 43%. Results for the four most influential inputs are included in Table 8.6. Consistent with the results derived using other approaches above, these results indicate that the contribution to variance in ingestion intake attributable to deposition velocity (Vd) is quite significant, with the other inputs contributing relatively little.

8.3.5. Nominal Range Sensitivity

Nominal range sensitivity (also known as local sensitivity analysis) computes the effect on model outputs exerted by varying each of the model inputs

TABLE 8.6 Partial Correlation Coefficients and Standardized Regression Coefficients for Inputs and Ingestion Intake (Case B)

Input	Partial Correlation Coefficient	Standardized Regression Coefficient
Vd	0.61	0.64
M_r	0.18	0.07
Q	0.20	0.05
Df	0.17	0.08

across its entire range of plausible values, while holding the other inputs at their nominal values. This technique addresses only a portion of the range of possible input values (Morgan and Henrion, 1990). It is described further in Chap. 3.

8.3.6. Probabilistic Sensitivity Analysis

Probabilistic sensitivity analysis involves running simulations in which different subsets of variable and/or uncertain inputs are assigned distributions, while all other inputs are set to their central values. Most commercially available Monte Carlo software includes an option to pursue this type of analysis. This approach allows one to identify the effect of interactions among logically-grouped sets of inputs and to infer whether certain groupings of inputs contribute most significantly to variance (or other key features) in the probabilistic results.

Probabilistic sensitivity analysis provides some interesting and unique information for analysts. In fact, this is the only technique discussed here that provides insight into how the third moment of the inputs affects the output. While correlation coefficients provide an indication of how the variance in a model output is affected by variance in a model input, probabilistic sensitivity analysis explores another potentially important interaction, the shift in the central tendency of the model results due to skewness in the shape of distributions for model inputs. It is possible that a positively skewed distribution for a particular model input may tend to increase the central value for dose or exposure, while at the same time other uncertain inputs may contribute more to the variance in model results. (Note: Inputs, for which a nonlinear relationship with the output exists, may also increase the central tendency of the output without contributing significantly to its variance.)

An analyst may first run a simulation in which skewed inputs are allowed to vary. Then he or she may run a second simulation in which skewed inputs are held fixed. Comparing the results of these two analyses, it is possible to identify subsets of inputs that have a profound influence on the central tendency in the output, versus subsets that influence variance in the output (Frey and Rubin, 1992a; Frey and Rhodes, 1996). This characteristic makes probabilistic sensitivity analysis a good method for validating the results of a statistical analysis. For example, when PCCs or SRCs have been used to identify inputs which are insignificant contributors to output variance, it is interesting to run an experiment in which those inputs are treated as point estimates. In this case, although the variance may not change much, the central tendency may change significantly. It is possible that some of the "insignificant" inputs must be added back into the analysis in order to preserve the central tendency of the output.

Another important use of probabilistic sensitivity analysis is to assess the relative importance of different sources of variability and uncertainty in two-dimensional analysis. Specifically, the analyst might first hold uncertain elements constant while allowing variable inputs to vary and then hold variable elements constant while allowing uncertain inputs to vary. Because model inputs may be both variable and uncertain, it is possible that sources of uncertainty in an input such as measurement error or random sampling error would be demobilized while simultaneously, sources of variability would remain active for the same input. An example of this type of probabilistic sensitivity analysis appears in Section 9.3.2 where a two-dimensional simulation of Case A is explored (see Figs. 9.13 through 9.15).

Frey and Rhodes (1996) illustrate how to characterize variance in correlations between inputs and outputs as a more general approach for identifying key contributors to both variability and uncertainty in two dimensional analyses.

8.3.7. Contribution to Variance

The contribution to variance of individual inputs in a strict sense refers to the fraction of the variance in the output contributed by each input in a model. The techniques outlined above only partially address this issue. However, the methods described above only assess the portion of the output variance which is explained by a linear (or monotonic in the case of rank regression) model. Kalagnanam et al. (1998) introduce a technique for contribution of variance assessment using weighted average importance sampling. In this approach, the results of a simulation are used repeatedly, giving different weights to each realization to estimate the effects of different distributions on input and output uncertainty. The variance in the output is estimated accounting for its dependence on the model inputs and the possible family of distributions which may be used to represent that input under uncertainty. In short, the expected reduction in the variance of a model output related to a percentage change in the variance of an uncertain input is calculated. Uncertainty importance is measured as the difference between the variance in the output, conditioned on the prior input distributions, and the mean value of the variance of the output, after reduction in input uncertainty.

8.4. SCREENING AND ITERATION

In the context of uncertainty analysis, the capability to identify and rank sources of uncertainty enables explicit consideration of the role of screening and iteration and its relationship to modeling. Rather than placing emphasis and resources on the development of detailed estimates of uncertainty for a comprehensive list of model inputs, one can start with preliminary characterizations. Then, statistical techniques may be used to identify which inputs contribute most to variance in model results and the process repeated with more refined distributions where warranted and plausible. Ultimately a stable set of sensitive inputs will be identified for additional attention with VOI techniques and possibly for future research efforts.

8.5. SUMMARY

Each of the techniques described above has strengths and weaknesses depending on the context in which it is applied. A summary of some advantages and disadvantages of each technique appears in Table 8.7.

TABLE 8.7 Summary of Techniques for Identifying Significant Contributors to Variance

Techniques	Advantages	Disadvantages/Cautions
Summary Statistics	• simple • can generate prior to simulation • use first order (Gaussian) approximation to relate sum of input variances to output variance	• first order approximation can introduce significant error if the input and output are related nonlinearly • looks at limited range of values of the inputs
Scatter Plot	• simple • allows visual inspection • identifies linear and nonlinear relationships, patterns, trends • good for hypothesis generation	• must generate after simulation • does not illuminate synergistic effects of two or more inputs on the output
Correlation Coefficient (Sample and Rank)	• simple to calculate • rank correlation estimates nonlinear, monotonic relationships	• must generate after simulation • rank correlation most useful if large differences in scale exist between inputs
Partial Correlation Coefficient Sample (PCC) Rank (PRCC)	• PCC measures linear relationship • PRCC useful when nonlinear relationships are present • PRCC useful in the case of large variations in scale and/or non-normal, skewed inputs or output • measure unique contribution of an input	• must generate after simulation • computationally intensive calculation • large variations in scale may distort PCC • not useful if relationship between inputs and outputs is nonmonotonic

(continued)

TABLE 8.7 (*Continued*)

Techniques	Advantages	Disadvantages/Cautions
Standardized Regression Coefficient	• estimates linear relationship between input and output without influence of scale • useful for examining the influence of input values across their entire plausible range • measure "shared" contribution of an input	• must generate after simulation • computationally intensive calculation • less informative when relationship between input and output is nonlinear
Nominal Range Sensitivity	• gauges the effect on output of changes from highest to lowest bounds of plausible values for each input	• addresses only a small portion of the possible space of the input values, because interactions among inputs are difficult to capture
Probabilistic Sensitivity Analysis	• simple • flexible, can be performed within commercial software packages • provides insight into how the third moment of the inputs may affect the central tendency of output • useful for validating results of a statistical analysis	• usefulness may depend directly on the relative individual contributions of the inputs and decisions about grouping • computationally intensive
Contribution to Variance	• does not rely on Taylor series approximations • can be used for complex, nonlinear, dynamic models	• computationally intensive calculation, however there are some weighting schemes available to ease this burden

CHAPTER 9

AN EXAMPLE OF PROBABILISTIC EXPOSURE ASSESSMENT IN THE COMMUNITY SURROUNDING NEW BEDFORD HARBOR

The following case example is provided to illustrate and support explanations of some of the issues and techniques presented in the preceding text. This example is not intended to be an exhaustive treatment of the variability and uncertainty in the potential human exposure from a contaminant source, but rather a hypothetical analysis of a narrowly defined case in terms of the exposure media, chemical compound of concern, location, and receptor characteristics. The purpose of this analysis (aside from the illustration of some of the concepts presented in this text) is to inform a decision-maker about exposure to a defined subpopulation during cleanup of a contaminated site. Further goals include understanding the extent of uncertainty and variability in exposure and identifying important contributors to overall variance in exposure. Decisions about design, monitoring, and implementation of future cleanups rely on this type of exposure information.

9.1. DEFINING THE EXPOSURE SCENARIO

The selected case is an assessment of human contact with PCBs originating from the New Bedford Harbor Superfund site via two exposure routes: i) inhalation; and ii) ingestion of locally grown produce (see Fig. 9.1). There are 209 individual chemical compounds known as PCBs; however, in this case the term refers to the sum of 59 of the more prevalent, persistent and toxic PCB congeners (Table 9.1). Many of the techniques and tools presented here could be adapted easily to assess human exposure to other contaminants. Although initial steps to clean up the harbor are underway, the underlying exposure assessment was based on historic data sets and information. Data collected during re-

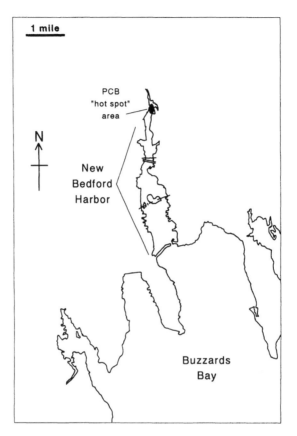

FIGURE 9.1 Schematic of New Bedford harbor.

cent cleanup operations lend themselves to illustrating many of the techniques presented in this text and to the hypothetical purpose of decision-making about community protection during the cleanup. Any specific individual's exposure is a function of the exact location where that person resides and spends time, and the manner in which the person comes into contact with the contaminant. Thus, before turning to the development of exposure model input distributions to represent variability and uncertainty, it is necessary to define the exposure in terms of averaging time, spatial range of interest, and some characteristics of the receptor.

9.1.1. Temporal Scope

First, we define the temporal scope of the analysis. Many exposure assessments are structured to support the information needs of health risk assessments.

TABLE 9.1 PCB Congeners Analyzed

IUPAC Number	Chlorobiphenyl Structure
6	2,3'
8	2,4'
18	2,2',5
16	2,2',3
31	2,4',5
28	2,4,4'
33	2'3,4
22	2,3,4'
52	2,2',5,5'
49	2,2',4,5'
44	2,2',3,5'
37	3,4,4'
74	2,4,4',5
70	2,3',4',5
66	2,3',4,4'
95	2,2',3,5',6
84	2,2',3,3',6
101	2,2',4,5,5'
99	2,2',4,4',5
97	2,2',3',4,5
87	2,2',3,4,5'
136	2,2',3,3',6,6'
110/77+	2,3,3',4',6/3,3',4,4'
151	2,2',3,5,5',6
135	2,2',3,3',5,6'
149	2,2',3,4',5',6
118	2,3',4,4',5
153	2,2',4,4',5,5'
105	2,3,3',4,4'
141	2,2',3,4,5,5'
138	2,2',3,4,4',5'
187	2,2',3,4',5,5',6
183	2,2',3,4,4',5',6
128	2,2',3,3',4,4'
167	2,3',4,4',5,5'
174	2,2',3,3',4,5,6'
177	2,2',3,3',4',5,6
171	2,2',3,3',4,4',6
156	2,3,3',4,4',5
157/201+	2,3,3',4,4',5'/2,2',3,3',4,5',6,6'
180	2,2',3,4,4',5,5'
170	2,2',3,3',4,4',5
199	2,2',3,3',4,5,5',6'
196/203+	2,2',3,3',4,4',5',6/2,2',3,4,4',5,5',6
189	2,3,3',4,4',5,5'
195	2,2',3,3',4,4',5,6
206	2,2',3,3',4,4',5,5',6

Thus it is necessary to define the relevant averaging time. For this illustration we assume an acute or subchronic health effect, such as endocrine disruption, is of concern, and consequently the exposure assessment is designed to produce an estimate of average daily dose, annually averaged. Since the period of remedial activity at the site is the period of interest, all field sampling carried out in support of distribution development coincided with the piloting and implementation of various stages of harbor dredging.

9.1.2. Spatial/Population Scope

Next we define the receptor(s) and spatial range of interest. We preserve a fairly narrow scope by defining the receptor of interest as a random adult female (21-years-old or older), who does not work outside the home, living within two or three miles of the New Bedford Harbor Superfund site (Fig. 9.1). Environmental samples and residential survey data used in this analysis were collected from this area.

9.1.3. Model Structure

Average daily dose (annually averaged) is computed according to the following models.

$$Dh = \frac{Ih}{Bw} \tag{9.1}$$

$$Dg = \frac{Ig}{Bw} \tag{9.2}$$

Dh = incremental average daily inhalation dose of PCBs to an adult female living within two or three miles of the New Bedford Harbor Superfund site, (mg/kg/day),

Dg = incremental average daily ingestion dose of PCBs to an adult female living within two or three miles of the New Bedford Harbor Superfund site, (mg/kg/day),

Bw = body weight of a randomly selected adult female, (kg),

Ih = average daily intake due to inhalation, annually averaged (mg/day),

Ig = average daily intake due to ingestion of local produce, annually averaged (mg/day).

Inhalation Intake (Ih)

$$Ih = Ir\{C_o(1 - Fi) + C_i(Fi)\} \tag{9.3}$$

where,

Ir = inhalation rate, annually averaged (m³/day),
C_o = concentration of PCBs in outdoor air, annually averaged (mg/m³),
C_i = concentration of PCBs in indoor air, annually averaged (mg/m³), and
Fi = fraction of time spent indoors, annually averaged (–).

Ingestion Intake via Locally Grown Produce (Ig)

$$Ig = \sum_{k=r,l,v} M_k C_k \tag{9.4}$$

where,

M_k = daily consumption rate of produce of type k, (g/day), and
C_k = concentration of PCBs in produce of produce of type k, annually averaged across items consumed in a one year period (mg/g).

Rather than consider each individual fruit and vegetable separately, produce is grouped into three categories based roughly on plant structure and uptake mechanisms: root (r), leafy (l), and vine (v) produce.

Average daily dose of PCBs via inhalation and ingestion of locally grown produce is computed according to Eqs. (9.1) through (9.4) using distributions that represent variability and uncertainty. Development of these distributions is outlined in Section 9.2. Data are summarized in Tables 9.2 through 9.5 and Figs. 9.2 through 9.11.

A more sophisticated exposure analysis estimating doses of PCBs also would account for the fraction of contaminant inhaled that actually is transported across the alveolar space or gastrointestinal tract into the bloodstream and ultimately to some target tissue. However, for the purpose of illustrating the techniques which are our focus, we restrict the model to the simple form (Eqs. (9.3) and (9.4)).

Three LHSs are used to explore exposure in the receptor population. In the first, a one-dimensional simulation of variability in exposure is carried out. In the second, uncertainty in the percentiles of variability in exposure is explored in an analysis which consists of 100 uncertainty realizations, each of which has 1000 variability realizations. In the third, variability in the percentiles of uncertainty is explored in an analysis consisting of 100 variability realizations, each of which has 1000 uncertainty realizations. All simulations are run in the Crystal Ball program. The analyses presented include treatments of the following sources of variance: uncertainty due to measurement error, uncertainty due to random sampling error (i.e., due to small sample sizes and our consequent lack

TABLE 8.2 Inputs to Dose Models

Input	Definition	Distribution Family	Parameters
Bw	body weight (kg)	lognormal	$m = 65$, $\sigma_{\ln(x)} = 0.2$
Ir	inhalation rate, annually averaged (m³/day)	lognormal	$m = 14$, $\sigma_{\ln(x)} = 0.2$
[b]M_k	daily consumption of locally grown produce, k, annually averaged (g/day)	lognormal	root produce: $m = 12.5$, $\sigma_{\ln(x)} = 0.6$ leafy produce: $m = 12.5$, $\sigma_{\ln(x)} = 0.6$ vine produce: $m = 35.0$, $\sigma_{\ln(x)} = 0.6$
C_i	concentration of PCBs in indoor air, annually averaged (mg/m³)	lognormal	parameters discussed in text
C_o	concentration of PCBs in outdoor air, annually averaged (mg/m³)	lognormal	parameters discussed in text
[a]Fi	fraction of time spent indoors, annually averaged (–)	beta	$\alpha_1 = 3.5$ $\alpha_2 = 0.876$ scale = 0.2
[b]C_k	concentration of PCBs in locally grown produce (mg/g)	lognormal	parameters discussed in text

[a]The fraction of time spent outdoors, is defined as $1 - Fi$.
[b]Individual produce types are subscripted k, i.e., root, vine, or leafy.

of knowledge about the true parameters of the distributions fitted to PCB concentrations in environmental media), and variability due to spatial, temporal (assuming annual averaging), and interindividual heterogeneity in the inputs. See Section 7.6 for general information about two-dimensional approaches.

Results of the one-dimensional analysis of variability appear in Table 9.6 and Fig. 9.12. Results of the two-dimensional analysis are presented in two formats. Cumulative distributions representing uncertainty in selected percentiles of (i.e., the 2.5th, 50th, 97.5th percentiles), as well as, the minimum and maximum simulated values, appear in Figs. 9.13 through 9.15. Cumulative distributions representing variability in selected percentiles of uncertainty (i.e., the 2.5th, 50th, 97.5th percentiles) as well as the minimum and maximum simulated values, appear in Fig. 9.16.

9.2. DISTRIBUTION DEVELOPMENT FOR EXPOSURE MODEL INPUTS

For clarity the exposure model inputs are organized into three groups: (i) concentration of PCBs in air and produce; (ii) human physical characteristics; and (iii) human rates of intake of air and produce.

Prior to the analysis all available information pertaining to each input of interest was gathered. Development of the distributions is based on field samples and homeowner survey data collected near this site as well as data found in the primary literature and assumptions about the defined population as discussed below.

9.2.1. Correlation

In the development of distributions it is important to consider the possibility of correlation among the inputs. Even in the simple models outlined above, there are four candidates to consider. First, an obvious correlation between fraction of time spent indoors and fraction of time spent outdoors was accounted for by the decision to simulate the former input (Fi) and calculate the latter $(1 - Fi)$ from the simulated value. Second, correlation is suspected between the concentrations of PCBs in indoor and outdoor air. These data were examined in detail, and no correlation was found (Vorhees et al., 1997). Third, correlation is possible among the concentrations of PCBs in various types of produce or by location. An analysis of concentration of PCBs by produce type and by location showed only a weak association (Cullen et al., 1996). Fourth, there is a possibility of correlation among the consumption rates of various types of produce. Given that the data available for this input are not in a form which would lend itself to such an assessment, we resolve to reassess whether it is important to include correlation in these inputs after an initial analysis indicates which of the inputs are contributing significantly to variance in the output. The alternative would be to speculate about the degree of correlation and run the analysis incorporating this guess. As discussed previously (Chap. 6), Smith *et al.* (1992) note that the seriousness of neglecting to specify correlation in probabilistic analysis depends on two things: (i) the degree to which inputs are correlated; and (ii) the contribution that each makes to the overall variance in output. Thus we reserve judgment about whether to act further in representing correlation until after the initial analysis.

9.2.2. Concentration of PCBs in Environmental Media

9.2.2.1. Concentration of PCBs in Locally Grown Produce.

Symbol: C_k

Definition:
The distribution of the daily average concentration of PCBs in produce grown in backyard gardens and small farms in the vicinity of New Bedford harbor and ingested by a random local resident during the one-year exposure period. The subscript k distinguishes root, vine, and leafy fruits/vegetables, and is coded r, v, and l respectively.

Units: mg/g

Distribution: based on Student's t, χ^2 and lognormal (See Chap. 4 for probability density functions and definitions.).

Parameters: The parameters are reported in two steps. For each produce type, PCB concentrations are sampled from a lognormal distribution. However, the parameters of this distribution are also assumed to be uncertain and governed by distributions. In the first step, two approaches are presented for generating distributions of the population parameters, mean and variance, of the distribution of PCB concentration for individual items of produce, (see Technical Basis and Approach for additional detail and definitions). In the first, more technically correct, approach the Student's t distribution is used to represent the distribution of the mean and the χ^2 distribution is used to represent the distribution of the variance (see Chaps. 4 and 5). In the second, simpler approach, the normal approximation is used for both. In the second step we present the equations for generating the distribution of the average PCB concentration by produce type accounting for the fact that multiple items are consumed by an individual receptor across a year.

References: Cullen *et al.*, 1996

Technical Basis: More than a dozen farms and gardens producing vegetables and fruit for local consumption are located within a few miles of the contaminated harbor site. Samples of this produce were collected by purchase at roadside stands or at the farms or gardens where they were grown, during two growing seasons (1992 and 1994). The sampling was timed to coincide to the extent possible with the piloting and implementation of harbor remediation. The samples were analyzed for PCBs in the trace level organics laboratory at the Harvard School of Public Health.

Our objective is to develop a distribution representing the average concentration of PCBs in each category of local produce ingested by receptors over the one-year exposure period. There are multiple sources of variability and uncertainty to be addressed in the analysis.

First, variability in concentration between samples is estimated based upon the distribution of concentration in the samples. It is assumed that an individual

TABLE 9.3 PCB Concentrations in Produce Samples[a]
(ng/g,wet weight)

Vine	Root	Leafy
0.08	0.16	0.09
0.10	0.09	0.35
0.06	0.11	0.05
0.08	0.31	0.28
0.07	0.14	0.28
0.04	0.19	0.18
0.25	0.10	0.21
0.11	0.11	0.25
0.13	0.06	0.30
0.11	0.47	
0.15	0.51	
0.10	0.47	
0.09	0.04	
0.06	0.17	
0.11	0.08	
0.05	0.16	
0.11	0.11	
	0.21	
	0.27	
	0.24	
	0.17	
	0.08	
	0.20	
	0.12	
	0.27	

[a]Cullen et al., 1996

item, e.g., one head of lettuce, is consumed across meals on several days by an entire household. Therefore, the concentration of PCBs in lettuce consumed by a receptor may not be independent across servings or days. A receptor is assumed to consume servings from 32 locally grown vine vegetables (e.g., squash or tomatoes), 9 locally grown root vegetables (e.g., potatoes or carrots), and 9 locally grown leafy vegetables (e.g., heads of lettuce) during one year (see Section 9.2.4.3 for a discussion of these assumptions). Thus, rather than make one random draw from the distribution of concentration variability to represent the entire year, we average multiple draws to generate the distribution of average concentration of PCBs in vine, root, or leafy produce respectively.

Second, there is uncertainty about concentration due to measurement error. This has been estimated at $\pm 10\%$ based on evaluation of duplicate samples which incorporate all sources of measurement error (Vorhees et al., 1997). Because these QA/QC samples serve to estimate error in batches of samples rather than individual samples, measurement error is represented for each medium in each simulation, but not for individual sample draws. Specifically, for each realization of uncertainty, a given estimate of measurement error is applied to all simulated values of variability for a single environmental medium. A further caveat about measurement error also pertains. In reality, measurement error and variability in concentration together make up the differences between sample concentrations measured by the laboratory. Measurement error is double-counted in our analysis, because the estimate of variability is based on differences between samples, which may include measurement error. However, measurement error is relatively small compared to variability, so this does not materially affect the results.

A third source of uncertainty about the concentration distribution is our lack of certainty about the extent to which the samples collected actually represent the contact medium. In other words, the produce samples collected may not be fully representative of those with which human receptors will come into contact. Some researchers recommend reducing the effective sample size to combat this problem (Brattin et al., 1996). Given our targeted collection of produce samples directly from the areas of interest and in the seasons of concern, there is no quantifiable lack of representativeness in these measured concentrations, so we do not reduce the effective sample size by an arbitrary amount.

Fourth, there is uncertainty due to random sampling error. In other words, efforts to fit distributions to small numbers of data points are complicated by the apparent consistency of the data with multiple plausible distributions.

There were several steps involved with developing the distributions to represent variability and uncertainty for each produce type. These same steps were carried out for the air concentration distribution development. As noted above, there are two approaches given for the distribution development. It is assumed that the distributions of the mean and standard deviation are statistically independent for both approaches. For the normal approximation this has been demonstrated (Laha and Rohatgi, 1979). While this assumption is not exact for the Student's t and χ^2 representations, it holds asymptotically for large sample size and is adequate for our purposes.

Approach

Normal and lognormal probability plots Figs. 9.2 through 9.4 were generated for the concentration measurements (Table 9.3) for visual inspection as a first step in the data exploration. The lognormal family of distributions was selected to represent variability in environmental contaminant concentrations, consistent with underlying theory and our inability to reject this family with such small sets of samples (see Chaps. 4 and 5 for detailed discussion of this issue). The population parameters, the mean and variance, of these distributions are uncertain due to limitations in sample size; however, their distributions can be generated based on the sample statistics, $\overline{\ln(x)}$ and $s_{\ln(x)}$.

Two approaches are presented for developing the distributions of these population parameters from the sample statistics, either of which can be used as input to the final step. Finally, the approach to the development of a distribution for the average concentration of PCBs across a specified number of items of produce is discussed.

1. Estimating Population Parameters
By Rigorous Approach

As discussed in Chap. 5, the distributions of mean and variance of a normal random variable, in this case the population mean concentration of PCBs in individual vegetables, are most rigorously based on the Student's t and χ^2 distributions respectively. We adapt this approach to a lognormally distributed random variable, applying it to the logarithms of the PCB concentration data.

The distribution of the population mean of the natural logarithm of concentration of PCBs for individual vegetables is:

$$\mu_{\ln x} \sim \overline{\ln(x)} + \frac{s_{\ln(x)}}{\sqrt{n}} t_{n-1} \quad \text{(after Eq. (5.24))} \quad (9.5)$$

where:

$\overline{\ln(x)}$ = the sample mean of the logtransformed concentration data,

$s_{\ln(x)}$ = the sample standard deviation of the logtransformed concentration data,

n = the number of samples measured for concentration,

t_{n-1} = a Student's t distribution with $n-1$ degrees of freedom.

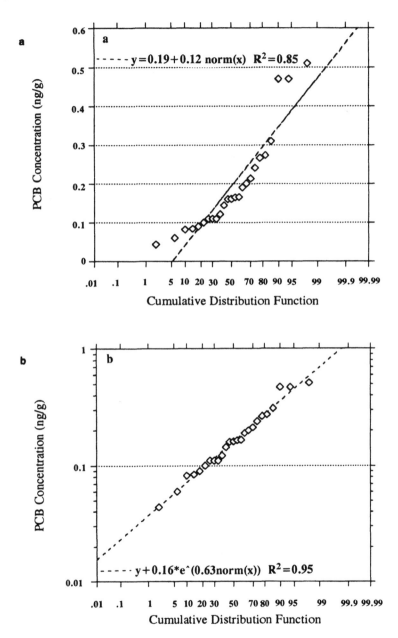

FIGURE 9.2 PCB Concentration measurements in root vegetables, a. normal fit b. lognormal fit.

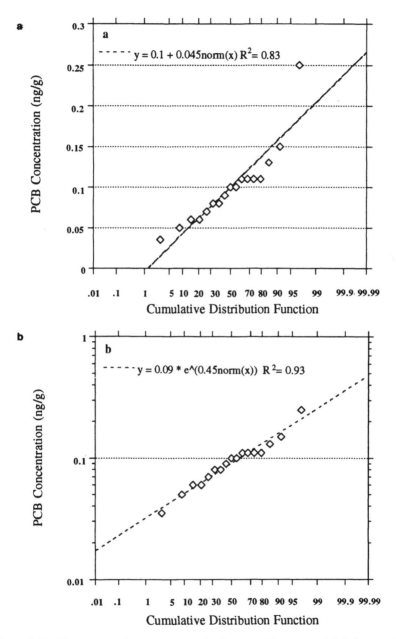

FIGURE 9.3 PCB Concentration measurements in vine vegetables, a. normal fit b. lognormal fit.

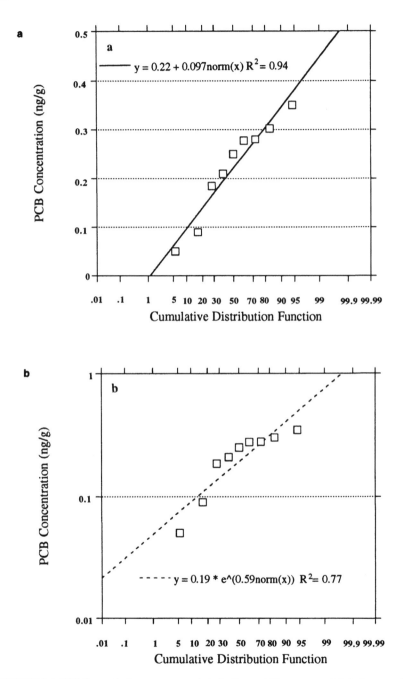

FIGURE 9.4 PCB Concentration measurements in leafy vegetables, a. normal fit b. lognormal fit.

The t distribution is simulated in the Crystal Ball and Excel programs as:

$$t_{n-1} \sim \frac{N(0,1)}{\sqrt{\dfrac{\chi^2_{n-1}}{n-1}}},$$

(9.6)

The χ^2_{n-1} distribution is simulated in the Crystal Ball program as a gamma distribution with scale = 2, and shape = $(n-1)/2$ (see Hastings and Peacock, 1974 for additional detail). $N(0,1)$ represents a normal distribution with mean equal to 0 and standard deviation equal to 1.

The distribution of the population standard deviation of the logarithm of concentration of PCBs for individual vegetables is:

$$\sigma_{(\ln x)} \sim \left[\frac{(n-1)s^2_{\ln x}}{\chi^2_{n-1}} \right]^{1/2} \quad \text{(after Eq. (5-26)).}$$

(9.7)

By Normal Approximation

In this approach we use normal distributions to represent the population mean and standard deviation of logtransformed concentration. Uncertainty about these parameters, due to small sample size, is reflected in their respective standard errors. The parameters and their standard errors are estimated based on the sample statistics or, alternatively, can be derived from lines fit to the concentration data probability plots.

Asymptotic approximations for the standard errors of the population parameters are:

$$\mu_{\ln(x)} \sim N\left(\overline{\ln(x)}, \frac{s_{\ln(x)}}{\sqrt{n}} \right),$$

(9.8)

$$\sigma_{\ln(x)} \sim N\left(s_{\ln(x)}, \frac{s_{\ln(x)}}{\sqrt{2n}} \right).$$

(9.9)

In practice these formulas produce accurate results for samples as small as 10 data points or fewer. Thus, with larger sample sizes the approximations might be used routinely; however, with sample sizes as small as 9 (e.g., leafy vegetables), we should properly investigate distributions based on the Student's t and the χ^2 distributions.

In Figs. 9.5 through 9.7, the distributions produced by the two approaches are compared in a one-dimensional simulation of variability and uncertainty com-

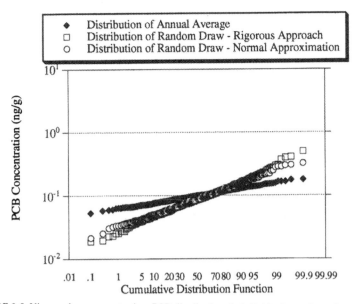

FIGURE 9.5 Vine produce concentration, PCB distributions in individual samples estimated using the rigorous approach and normal approximation for parameter estimation, and distribution of annual average PCB concentration.

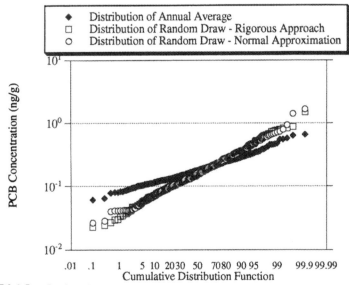

FIGURE 9.6 Root Produce Concentration, PCB distributions in individual samples estimated using the rigorous approach and normal approximation for parameter estimation, and distribution of annual average PCB concentration.

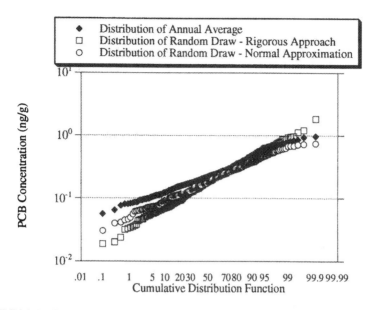

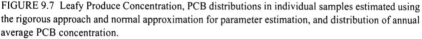

FIGURE 9.7 Leafy Produce Concentration, PCB distributions in individual samples estimated using the rigorous approach and normal approximation for parameter estimation, and distribution of annual average PCB concentration.

bined. (The annual average distributions in these figures are discussed in the next step.) Although the sample sizes for the three produce types range from 9 samples of leafy vegetables to 25 samples of root vegetables, the two approaches produce remarkably consistent results even at the 99th percentile. The asymptotic normal approximation appears to perform well for these datasets, although it does perform better for a sample size of 25 (Fig. 9.6) than for a sample size of 9 (Fig. 9.7).

Sample Statistics for Parameter Estimation

$$\text{Vine Produce: } n = 17,\ \overline{\ln(x)} = -2.4,\ s_{\ln(x)} = 0.43$$

$$\text{Root Produce: } n = 25,\ \overline{\ln(x)} = -1.8,\ s_{\ln(x)} = 0.63$$

$$\text{Leafy Produce: } n = 9,\ \overline{\ln(x)} = -1.65,\ s_{\ln(x)} = 0.59$$

2. Estimating the Parameters of the Distribution of the Average Concentration of PCBs Across Multiple Produce Items, <x>

As discussed, our overall objective is to generate a distribution for the average concentration of PCBs in vine, root, or leafy produce across a year of exposure. Thus a distribution representing the average,$<x>$, of multiple draws from the distributions of PCB concentration is required. The distribution of $<x>$ has the same mean as that of individual produce samples, reduced by a factor of N (the number of draws). The distribution of $<x>$ is represented as a lognormal to ensure positive concentrations. For large N it is actually closer to a normal, because it represents the average of a set of random variables, but the difference between the two distributions becomes negligible in this case.

Given the parameters $\mu_{\ln(x)}$ and $\sigma_{\ln(x)}$ generated according to Eqs. (9.5) to (9.7) or (9.8) and (9.9) we calculate the relevant parameters $\mu_{\ln<x>}$ and $\sigma_{\ln<x>}$ according to Eqs. (9.10) through (9.13).

The imposing formulae for the parameters of $<x>$ result from the derivation of a lognormal distribution whose mean is the same as that of x, but whose variance is $1/N$th as large as that of x.

$$\mu_{\ln<x>} = \ln\left[\frac{\mu_x}{\sqrt{1 + \frac{\sigma_x^2/N}{\mu_x^2}}}\right] \qquad (9.10)$$

$$\sigma_{\ln<x>} = \sqrt{\ln\left(1 + \frac{\sigma_x^2/N}{\mu_x^2}\right)} \qquad (9.11)$$

where

$$\mu_x = \exp\left(\mu_{\ln(x)} + \frac{\sigma_{\ln(x)}^2}{2}\right) \qquad (9.12)$$

$$\sigma_x = \exp(\mu_{\ln(x)}\sqrt{(\exp(\sigma_{\ln(x)}^2))(\exp(\sigma_{\ln(x)}^2) - 1)} \qquad (9.13)$$

N = the number of items of one produce type to which a receptor is exposed during one year.

The annual average distributions in Figs. 9.5 to 9.7 (which were generated using distributions of $\mu_{\ln(x)}$ and $\sigma_{\ln(x)}$ from the rigorous approach) show the expected reduction in variability due to averaging, in which the range of typical variability is $1/N$th as large as for individual samples.

9.2.2.2. Concentration of PCBs in Indoor and Outdoor Air.
Symbol: C_i, C_o

Definition: The concentration of PCBs in indoor and outdoor air at a receptor residence within 2–3 miles of New Bedford Harbor, annually averaged.

Units: mg/m^3

Distribution: based on Student's t, χ^2 and lognormal (See Section 9.2.2.1 for a more complete development of the two approaches and Chap. 4 for probability density functions and definitions.).

Parameters:
The approach is similar to that described above for developing concentration distributions for produce (Section 9.2.2.1). Sampling distributions for the parameters of the normal distributions for the natural logarithms of the concentration values are based on the Student's t distribution for the mean and χ^2 distribution for the variance. Indoor air and outdoor air are handled slightly differently as discussed in the Technical Basis and Approach below.

References: Vorhees et al., 1997

Technical Basis:
Air samples were collected indoors and outdoors from residences and neighborhoods within 2–3 miles of the harbor between April 1994 and April 1995, i.e., during the initial dredging period (Vorhees et al., 1997). These serve as the basis for the distributions representing these inputs. There are multiple sources of variability and uncertainty to be considered during distribution development for concentration of PCBs in air.

First, there is variability in concentration among samples due to spatial and temporal heterogeneity.

For indoor air, we estimate spatial variability based upon the distribution of concentration measurements in our samples (Table 9.4). For simplicity we assume no temporal variability for indoor air concentration. In other words, it is assumed that a receptor inhales the air in one particular house for the portion of the exposure year spent indoors. This assumption seems reasonable since the receptor definition excludes those working outside the home, and variability between houses is greater than temporal variability within a home (Vorhees et al., 1997). At worst this assumption leads to a slight overestimate of the spatial variability in concentration.

We can not assume that a receptor inhales outdoor air from the same location for the entire year and that PCB concentration in that air does not change over the year. Outdoor exposure occurs not only in one's own yard, but also in parks, and other places around town. In addition, concentration of PCBs in outdoor air varies with temperature and wind direction. We address this issue by

TABLE 9.4 PCB Concentrations
in Indoor Air Samples (ng/m³)

PCB Concentration ng/m³
17.83
17.01
8.47
17.1
20.05
51.82
7.54
48.88
19.02
13.56
10.57
11.52
17.11
12.45
7.61
12.72
42.18
11.75
32.96

(adapted as a subset from Vorhees *et al.*, 1997)

assuming that over the year all receptors experience essentially the same average PCB concentration in outdoor air and that weather patterns occur in roughly five-day-cycles in New England. Since sampling occurred fairly uniformly throughout the seasons, we develop an annual average distribution by averaging N = 365/5 = 73 independent draws from the outdoor air PCB concentration distribution based on the set of measurements (Table 9.5). See Section 6.3 for a discussion of the decision to assign equal weight to all outdoor air samples regardless of the season in which collection occurred.

As with produce samples, there are several sources of potential uncertainty about concentration of PCBs in air. Uncertainty due to measurement error has been estimated at ±10% based on evaluation of duplicate samples and incorporate all sources of measurement error (Vorhees et al., 1997). Uncertainty about the representativeness of the samples relative to the contact medium for human exposure is not quantifiable given our sampling scheme and approach. Uncertainty due to random sampling error is incorporated in the analysis as discussed in Section 9.2.2.1. See Chap. 5 for a detailed treatment of random sampling error.

There were several steps involved with developing the distributions to represent variability and uncertainty which are described above with respect to distribution development for concentration in produce. To that discussion we add a few details particular to the air samples.

Normal and lognormal probability plots of the concentration data appear in Figs. 9.8 and 9.9 for visual inspection. The selection of the lognormal to represent these distributions, as with produce, is more than a statistical fitting exercise (see Chap. 5). Rather, it is compelled by theoretical work on the underlying physics of dilution processes in air (Ott, 1990).

Approach

Indoor Air

The distribution of concentration of PCBs in indoor air was developed based on the data in Table 9.4. We note that the annual average distribution of PCBs in indoor air is the distribution of the measurements (see Technical Basis).

The initial steps laid out in Section 9.2.2.1 were followed for the two approaches to parameter estimation for the Lognormal concentration distributions, based on the sample statistics:

$$\text{Indoor air: } n = 19, \overline{\ln(x)} = -11.0, s_{\ln(x)} = 0.59.$$

Outdoor Air

The parameters of the lognormal distribution for concentration of PCBs in outdoor air were calculated using data from Table 9.5. The complete approach outlined in Section 9.2.2.1 was used for constructing a distribution for annually averaged concentration assuming 73 weather cycles of five days each occurring across the year (see Technical Basis). The sample statistics are:

$$\text{Outdoor air: } n = 30, \overline{\ln(x)} = -12.37, s_{\ln(x)} = 1.56.$$

As with the produce concentration distributions, the asymptotic normal approximation to the Student's t and χ^2 distributions for the population mean and standard deviation performs well even out in the tails for indoor and outdoor air concentration, and there is a large reduction in PCB concentration variability in outdoor air after averaging over many weather cycles (Figs. 9.10 and 9.11).

Parameters of the Distribution of the Average Concentration of PCBs in Outdoor Air, <x>, Across Multiple Weather Cycles

The parameters of the distribution of average concentration are given in Eqs. (9.10) through (9.13), where N is the number of weather cycles during one year, i.e., 73.

TABLE 9.5 PCB Concentrations in Outdoor Air Samples (ng/m³)

PCB Concentration ng/m³	Season[a]
15	1
11	1
7.6	1
18	1
36	1
16	1
14	1
31	1
21	1
0.94	2
0.45	2
0.64	2
0.82	2
0.82	2
0.94	2
1.2	2
21	3
1.3	3
1.1	3
5.6	3
0.95	3
0.85	3
11.0	3
0.45	3
0.73	3
3.5	3
11	3
56	3
26	3
10	3

[a]Season 1 (June, July, August, September) $n = 9$
Season 2 (December, January, February) $n = 7$
Season 3 (October, November, March, April, May) $n = 14$
(adapted as a subset from Vorhees et al., 1997)

9.2.3. Human Physical Characteristics

9.2.3.1. Body Weight.

Symbol: Bw

Definition: Body weight is the weight of an adult female selected at random from the population.

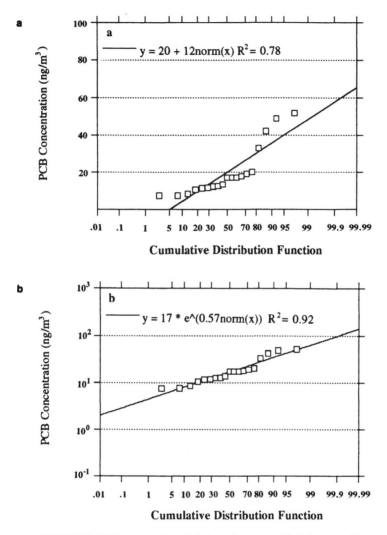

FIGURE 9.8 PCB concentrations in indoor air, a. normal fit b. lognormal fit.

Units: kg

Distribution: lognormal (See Section 4.3.1.1 for probability density function and definitions.)

Parameters: m = median = 65, $\sigma_{\ln(x)}$ = 0.2

References: Brainard and Burmaster, 1992

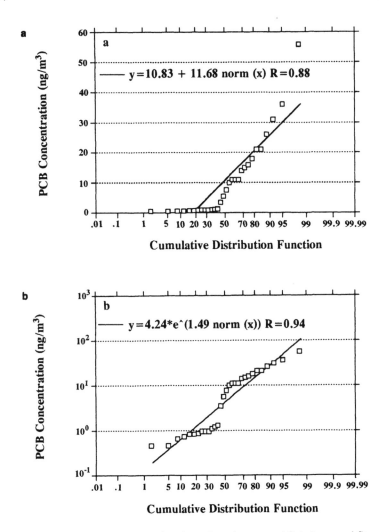

FIGURE 9.9 PCB concentrations in outdoor air, a. normal fit b. lognormal fit.

Technical Basis:

The present example is concerned with exposure to PCBs by an adult female selected at random from the defined area and population, thus it is appropriate to assign a distribution whose spread reflects interindividual variability. A great deal of interindividual variability distinguishes human body weight; however, for a single individual body weight is assumed to be fairly consistent over the course of adulthood. Using information collected under the second National Health and Nutrition Examination Survey (NHANES II), which included

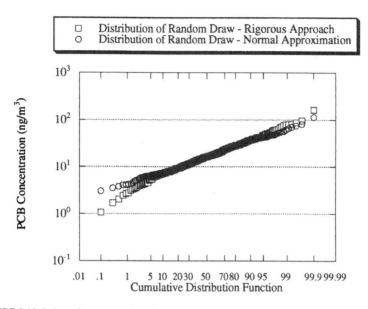

FIGURE 9.10 Indoor air concentration distribution: Comparison of rigorous approach and normal approximation.

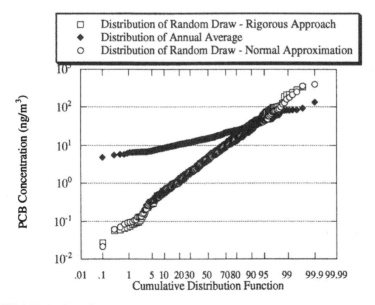

FIGURE 9.11 Outdoor air concentration distribution: PCB distributions in individual samples estimated using the rigorous approach and normal approximation for parameter estimation, and distribution of annual average PCB concentration.

a survey of 6588 women selected to represent adult females in the United States population as a whole, a lognormal distribution of body weight was developed by Brainard and Burmaster (1992). This distribution is adopted for use in the present analysis under the assumption that the New Bedford population of interest has the same body weight distribution as the corresponding population (women) in the NHANES survey. Uncertainty in the parameters of the body weight distribution is not tackled in this analysis on the assumption that variability between the body weights of individuals is far greater than errors produced by the scales used to weigh them, or possible misrepresentativeness.

9.2.4. Inputs Related to Intake and Consumption Rates

9.2.4.1. Inhalation Rate,

Symbol: *Ir*

Definition: The inhalation rate is the average daily volume of air inhaled by an individual.

Units: m^3/d

Distribution: lognormal (See Section 4.3.1.1 for probability density function and definitions.)

Parameters: m = median = 14, $\sigma_{\ln(x)}$ = 0.2

References: USEPA, 1995; Layton, 1993.

Technical Basis:
There are no available measurements of inhalation rate for the general population. This lack of data leaves the analyst grappling with uncertainty about the appropriate choice of distribution family to represent variability between individuals, as well as uncertainty related to the conversion of data relating to other physiological characteristics, such as measures of metabolic activity, and estimates of inhalation rate. We assume a lognormal distribution on the basis of observation that other human physiological characteristics are fit reasonably well by this distribution family. Given that the hypothetical individual in question is drawn from a heterogeneous population, it is necessary to reflect inter-individual variability in the geometric standard deviation of the lognormal, while the median should represent the central tendency for the population of adult females as a whole, not any single individual. Based on research on metabolic demand for oxygen by Layton (1993) and its relationship to inhalation rate, the USEPA (1995) recommends a median for inhalation rate for adults of 14 m^3/day. USEPA further suggests that 20 m^3/d represents an upper percentile estimate. By assigning the 95th percentile to 20 m^3/d, a geometric standard deviation of 1.2 is calculated (i.e., gsd $\cong$ (95th percentile/50th percentile)$^{0.5}$, or $(20/14)^{0.5}$). Un-

certainty in the parameters of the inhalation rate distribution is not tackled in this analysis although it would be interesting as a next step.

Due to the lack of directly relevant information it is difficult to know if our estimate of the median inhalation rate is appropriate or in particular, it is appropriate for adult females as a subpopulation. In some exposure assessments the results could be significantly compromised by this lack; however, in this analysis the inhalation rate will later be shown to be an insignificant contributor to overall variance, shadowed by the importance of the concentration distributions.

9.2.4.2. Fraction of Time Spent Indoors.

Symbol: Fi

Definition: The fraction of time spent indoors daily by a random adult female not working outside the home, in the Greater New Bedford area, annually averaged.

Units: (–)

Distribution: beta (See Section 4.3.1.5 for information about this distribution)

Parameters:

$$\alpha_1 = 3.5$$
$$\alpha_2 = 0.876$$

based on: lower bound $= \lambda = 0.8$
upper bound $= \upsilon = 1.0$
most likely value $= \mu = 0.96$

References: Sexton and Ryan, 1987

Technical Basis:
The distribution of the fraction of time spent indoors was based on a combination of published work from Sexton and Ryan (1987) and subjective judgment drawing on knowledge about the climate in the area of interest and the population. Sexton and Ryan summarized data on fraction of time spent indoors from two studies from the 1970s. They found that individuals on average spend the vast majority of their time indoors or in transit, while less than 1 hour per day was spent outside. Given the age of the studies and the lack of current or local information on time-activity patterns we decided to develop a beta distribution for variability in this input using maximum entropy inference (see Section 6.2). Uncertainty was not characterized for this input.

The maximum entropy approach requires estimates of the lower bound, upper bound and most likely value, to fit the beta distribution. The upper limit is assumed to be 1.0 (i.e., 100% of time is spent indoors), since it is possible for an individual in the target population of interest to spend no time outdoors in the local area, only indoors and in transit to other locations. The lower limit and best estimate of the distribution are based on knowledge of the climate and seasons of the New Bedford area.

The lower limit, averaged over the year, assumes a person spends as much time as possible outside. In the warmer months a person who is interested in gardening, boating or other outdoor activities or who has a porch or lawn furniture may spend much of the day outside while in the colder weather these opportunities would be limited. In this case we assume that in June, July, August, and September, the four warmest months, that the individual spends 9 hours outside and 15 hours inside each day. In the spring and fall (October, November, March, April, and May) the individual is assumed to spend 4 hours outside and 20 hours inside each day. In the coldest months of the year (December, January, and February) the individual is assumed to spend 1 hour per day outside and 23 hours inside. This division of time between indoors and outdoors is assumed to be the daily average across each season. For example, this averaging could account for occasional 16-hour-summer days in the yard, as well as occasional summer days spent entirely in the house. Averaged over the year the outdoors person spends time indoors as follows:

$$\left(\frac{15 \text{ hours}}{24 \text{ hours}} \times \frac{4 \text{ months}}{12 \text{ months}}\right) + \left(\frac{20 \text{ hours}}{24 \text{ hours}} \times \frac{5 \text{ months}}{12 \text{ months}}\right) + \left(\frac{23 \text{ hours}}{24 \text{ hours}} \times \frac{3 \text{ months}}{12 \text{ months}}\right)$$

$$= 0.80$$

i.e., approximately 80% of the time is spent indoors and 20% is spent outdoors.

The most likely value is generated based on personal judgment about fairly standard durations of time spent outdoors in a region having seasons. In the warmer months a person is assumed to spend part of the day pursuing an outdoor activity while in the colder weather outdoor time is assumed to be limited to walking to and from the house and car. In June, July, August, and September, the four warmest months, a hypothetical individual is assumed to spend an average of 2 hours outside and 22 hours inside each day. In the spring and fall (October, November, March, April, and May) the individual is assumed to spend half an hour (0.5) outside and 23.5 hours inside each day. In the coldest months of the year (December, January, and February) the individual is assumed to spend a quarter of an hour (15 minutes) per day outside and 23.75 hours inside. Averaged over the year the most likely value for time spent indoors is:

$$\left(\frac{22 \text{ hours}}{24 \text{ hours}} \times \frac{4 \text{ months}}{12 \text{ months}} \right) + \left(\frac{23.5 \text{ hours}}{24 \text{ hours}} \times \frac{5 \text{ months}}{12 \text{ months}} \right) + \left(\frac{23.75 \text{ hours}}{24 \text{ hours}} \times \frac{3 \text{ months}}{12 \text{ months}} \right)$$

$$= 0.96$$

Thus, approximately 96% of the time is spent indoors and 4% is spent outdoors. This assumption is approximately equal to reported values found in two time-activity data sets summarized by Sexton and Ryan (1987). See Section 6.2 for the fitting of parameters by maximum entropy inference.

9.2.4.3. Consumption of home/locally grown produce

Symbol: M_k

Definition: the daily average mass consumed of specific individual produce types (distinguished by k = root, leafy, or vine), annually averaged.

Units: g/d

Distribution: lognormal (See Section 4.3.1.1 for probability density function and definitions.)

Parameters:

> Root Produce
> m = median = 12.5, $\sigma_{\ln(x)} = 0.6$
>
> Leafy Produce
> m = median = 12.5, $\sigma_{\ln(x)} = 0.6$
>
> Vine Produce
> m = median = 35, $\sigma_{\ln(x)} = 0.6$

References: USDA, 1983; Cullen et al., 1996

Technical Basis:
Great diversity in diet and specifically, consumption, of locally grown produce, exists by geographic region, ethnic background, age, and economic status in the United States. Through extensive surveying of the Greater New Bedford area we identified at least a dozen local farms and gardens which sell produce grown on the premises, thus establishing diet as a plausible exposure pathway (Cullen *et al.*, 1996). Survey data collected as part of a larger project revealed that many residents consume locally grown produce consistent with this assumption (Vorhees *et al.*, 1997). In the absence of specific data regarding dietary preferences of residents of Greater New Bedford we estimate consumption rates based

on the Department of Agriculture's national dietary survey of 1977–1978 (USDA, 1983). Typical total daily consumption of vegetables is 200 g/day according to EPA, while for fruits 140 g/day has been estimated. Of this total the national homegrown/locally grown fraction is 4–75%, depending on type of vegetable, with a median of 25%, and for fruits is 9–33%, with a median of 20%. Estimates of the locally produced mass of vegetables and fruits consumed daily were 50 g/day median vegetable consumption and 30 g/day median fruit consumption.

To develop distributions we first plotted food survey data from USDA (1983) on log-probability paper based on our reckoning that like most human characteristics, consumption rate would vary as a lognormal distribution. Visual inspection of the plots satisfied us that this was not unfounded. Based on the observation that the ratios of the 95th/5th percentiles ($\sim$gsd^4) for consumption rate reported by produce item were generally approximately 10, we calculated a geometric standard deviation of 1.8 for consumption rate, (i.e., $10^{1/4} = 1.8$ and $\sigma_{\ln(x)} = 0.6$). Next, the daily intake of produce was divided into vine, root, and leafy categories, and median consumption estimates were calculated from the data. This calculation yielded estimates of 35 g/day vine produce, 12.5 g/day leafy produce, and 12.5 g/day root produce, based on the USDA database.

The development of a distribution representing PCB concentration in produce requires that we average across the number of items from each produce category consumed by one receptor during one year (see Section 9.2.2.1). The number of items can be calculated using the consumption rates and an assumption about the mass of one item. For example, if the edible portion of a potato (representative root vegetable) weighs about 500 g, and a receptor consumes about 12.5 g per day for a year, then they consume approximately 9 locally grown root vegetables per year, i.e., N = (1 potato/500g)(365 days/year)(12.5 g/day) = 9 potatoes. We also assume that the edible portion of a head of lettuce weighs 500 g and thus receptors also are assumed to consume approximately 9 locally grown leafy vegetables per year. Similarly, if the edible portion of a tomato or squash (representative vine vegetables) weighs about 400 g, and a receptor consumes about 35 g per day for a year, they consume approximately 32 locally grown vine vegetables per year. A more precise treatment would account for the fact that portions of individual items are likely to be consumed by other household members as well.

The next logical step in distribution development would consist of estimating uncertainty due to the limitations of the design of USDA dietary surveys; however, that step is not included in this analysis. This uncertainty is a result of the use of short term dietary surveys to estimate longer term consumption rates. Obtaining longer term consumption rate data or multiple short surveys would allow us to compare the "apparent" variability from each for consistency. Differences in apparent variability could be interpreted as possible indicators of measurement error. Unfortunately such information is also quite scarce. For some exposure analyses measurement error in this input might affect the overall results significantly; however, we find that inhalation dose is approximately a factor of 10 larger than ingestion dose and thus it is of less concern for this analysis.

9.2.5. Setup of Analysis

Three types of simulations were carried out to explore the variability and uncertainty in exposure for this case. The first was a one-dimensional simulation of variability, intended to provide insight into the model. (An additional one-dimensional simulation of variability and uncertainty was run to examine the relative importance of the individual contributors to overall variance in exposure (see Chap. 8)). The second type of simulation generated distributions of uncertainty around specified percentiles of variability (2.5, 50, and 97.5th percentiles). This type of analysis is often of interest in decision-making since it provides information about the exposures faced by individuals at upper end percentiles of variability, and our uncertainty about the magnitudes of those exposures. The third type of simulation, closely related to the second, generated distributions of variability around specified percentiles of uncertainty (2.5, 50, and 97.5th percentiles). This type of analysis provides insight into the mathematical properties of uncertainty frequency distributions and gives decision-makers information about the sensitivity of their decision to uncertainty. An example of a two-dimensional simulation of variability and uncertainty is given in Section 7.6.6, in which uncertainty and variability are simultaneously considered for all percentiles.

These simulations take varying degrees of time and effort to set up. The one-dimensional simulation is a quick way to check models and inputs and the mechanics of the analysis. The two-dimensional simulations are more complex. The results in this text were generated using the steps given here. These steps are for making CDFs of uncertainty in specific variability percentiles. Simply reverse the roles of variabilty and uncertainty to produce the CDFs of variability on selected percentiles of uncertainty.

1. Set up Model Inputs for PCB Concentration: Set up 100 identical rows in a spreadsheet containing cells for environmental concentration. This analysis used an Excel/Crystal Ball framework although others perform similarly. For the more rigorous approach, the population mean is distributed based on a t distribution as described in Eqs. (9.5) and (9.6), while the population standard deviation is distributed based on a χ^2 distribution, as described in Eq. (9.7). For the approach based on the normal approximation, variability in concentration is handled by a standard normal variate, $Z \sim N(0,1)$, with appropriate transformations for location and scale. Random sampling error is handled by assigning normal distributions to the population parameters. During simulation random draws from the $N(0,1)$ are combined with the μ and σ drawn from the distributions of these parameters to form a distribution of values of concentration $Z\sigma + \mu$ reflecting variability and random sampling error for concentration. For both approaches measurement error is handled by simulating a $\varepsilon \sim N(0,0.1)$ for each environmental medium. During iterations in the variability dimension $1 + \varepsilon$ is multiplied by each simulated concentration.

2. Set up Model Inputs for Human Physical, Physiological and Behavioral Characteristics: For these inputs, distributions of variability are entered directly into the cells of the spreadsheet.

3. Set up Forecast cells: Add cells to each row of the spreadsheet for the model formulas for Inhalation and Ingestion Dose (Eqs. (9.1 and (9.2)), as well as the sum of these which we label Total Dose (although this is for convenience and not strictly true since it does not include other possible exposure pathways, such as dermal absorption).

4. Simulation of Uncertainty: Make one random draw from the distribution for each uncertain quantity. This will result in each of the 100 rows having distinct values for the uncertain inputs. Note that it is the concentrations of PCBs in environmental media that are being simulated with an uncertain component in this analysis (i.e., measurement error and the components used to generate the population parameters for concentration, such as the Student's t or χ^2 variates). For the concentration of PCBs in indoor air the estimates of the population parameters are frozen before continuing. For outdoor air and produce, the population parameter estimates are used to calculate the parameter estimates for the distribution of concentration across multiple draws, and then these quantities are frozen before continuing.

5. Simulation of Variability: Make 1000 random draws for each variable quantity, thus making a distribution of variability for each row.

6. Exploring Percentiles of Interest: Extract either the full set of results or simply the percentiles of interest. Plot these as CDFs of selected variability percentiles. The 100 rows will each produce a single value of the selected percentile of exposure from their variability distributions. These 100 values encompass the uncertainty in that variability percentile.

7. Computational Demands: On a PowerMac with 65 MB of RAM using the Excel and CrystalBall programs, steps 4 through 6 take approximately 15 minutes to execute.

TABLE 9.6 Results of the One-dimensional Simulation of Variability

Percentiles	Dh Inhalation Dose (mg/kg/day)	Dg Ingestion Dose (mg/kg/day)	Total Dose (mg/kg/day)
2.5	1.1×10^{-6}	4.7×10^{-8}	1.2×10^{-6}
25	2.4×10^{-6}	9.7×10^{-8}	2.6×10^{-6}
50	3.5×10^{-6}	1.4×10^{-7}	3.7×10^{-6}
75	5.3×10^{-6}	2.0×10^{-7}	5.4×10^{-6}
97.5	1.1×10^{-5}	4.1×10^{-7}	1.1×10^{-5}

9.3. RESULTS

9.3.1. One-Dimensional Analysis of Variability

Table 9.6 contains the numerical results of the one-dimensional simulation of variability. We observe that inhalation dose is twenty-fold higher than ingestion dose.

Fig. 9.12 contains the graphical results of the one-dimensional simulation. These results indicate that the range of variability encompasses slightly more than one order of magnitude.

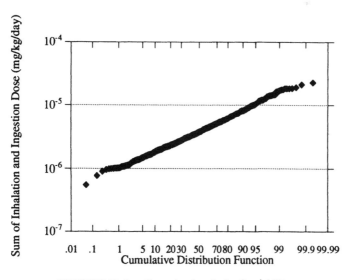

FIGURE 9.12 One-dimensional analysis of variability.

9.3.2. Two-Dimensional Analysis of Variability and Uncertainty

Figs. 9.13 through 9.15 contain the results of three analyses of the overall variance in dose (sum of inhalation and ingestion): (a) variability, and uncertainty due to measurement error; (b) variability, and uncertainty due to random sampling error, i.e., a lack of knowledge about the true parameters of the fitted distributions for PCB concentrations in environmental media with no measurement error; and (c) variability, and both sources of uncertainty.

In Fig. 9.15 we observe that variability, e.g., indicated by the degree of spread across the 2.5th, 50th, and 97.5th percentiles, is greater in magnitude than uncertainty about any of those percentiles, indicated by the range covered by any single line of plotted points. Note that on an absolute basis, uncertainty increases at the upper percentiles, a result of the effect of our decreasing ability to represent the concentration distributions as we move out into the tails. Uncertainty therefore is greater at both tails than at the median of the cumulative distribution. Comparing Figs. 9.13 and 9.14 we observe that the uncertainty contribution of the random sampling error outweighs the impact of the measurement error, i.e., the percentile plots are steeper when random sampling error is incorporated without measurement error, than when the reverse is done. Measurement error

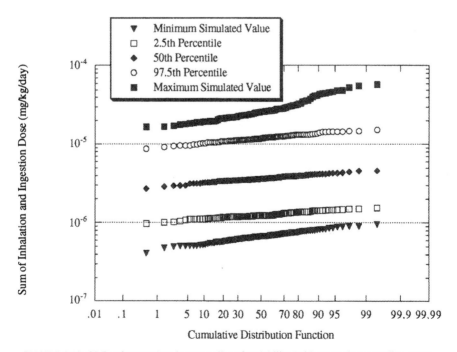

FIGURE 9.13 CDFs of uncertainty in percentiles of variability (without random sampling error).

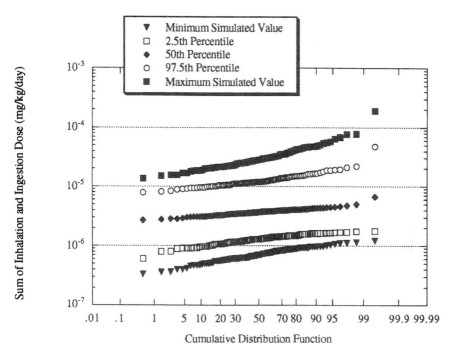

FIGURE 9.14 CDFs of uncertainty in percentiles of variability (without measurement error).

is not a strongly influential source of variance, thus our decision to base estimates of variability in annually averaged concentration on the measured PCB concentrations in produce and air, rather than pursue deconvolution of measurement error and variability first, is not compromising our conclusions.

Another way to look at the relative magnitudes in variability and uncertainty is by comparing the spread of the percentiles of uncertainty with the range of the CDFs of variability in Fig. 9.16. We observe a fairly tight grouping of the uncertainty percentiles (the vertical height of the "band" of plots) relative to the spread exhibited in the variability CDFs (the vertical distance traveled by a single percentile plot). For example our 95% confidence interval about exposure for the 95th percentile of population variability is equal to the vertical distance between the 2.5th percentile plot and the 97.5th percentile plot at the 95th percentile of the cumulative distribution (the horizontal axis). This confidence interval is less than an order of magnitude. From this figure we can see that uncertainty is greater, on a relative basis, at the extremes, than near the central percentiles of variability, since the band of plots flares at each end.

In this case we have explored dose for a specifically defined receptor, randomly chosen. Variability is found to contribute more significantly than uncer-

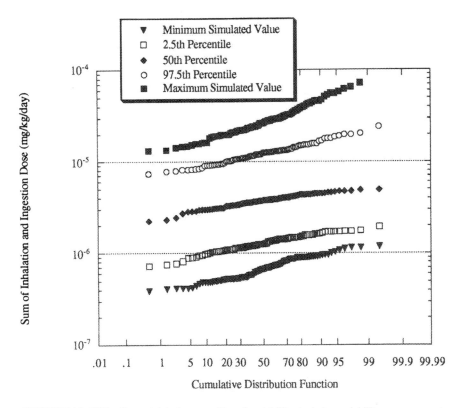

FIGURE 9.15 CDFs of uncertainty in percentiles of variability, includes variability, measurement error, and random sampling error.

tainty to overall variance. It is worth noting that if the decision end point had involved an entire population (e.g., a decision about population risk), the variability across individuals would have been integrated out leaving just uncertainty.

9.3.3. Identification of Key Contributors to Variability in Output

Multiple methods for identifying significant contributors to overall variance in inhalation and ingestion dose are applied to the one-dimensional results for variability and uncertainty combined (see Chap. 8). As reported there, by far the most influential input with regards to variance in inhalation dose is the PCB concentration in indoor air (C_i). The most influential inputs to variance in ingestion dose are the concentration of PCBs in vine produce (C_v) and the rate of consumption of vine produce (M_v).

As discussed in Section 9.3.2, probabilistic sensitivity analysis was used to compare the relative magnitudes of the different sources of uncertainty and vari-

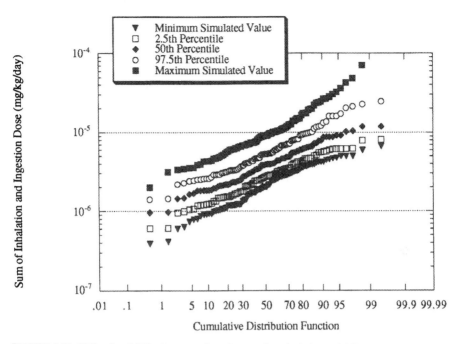

FIGURE 9.16 CDFs of variability in percentiles of uncertainty, includes variability, measurement error and random sampling error.

ability. We observe that variability plays a greater role than uncertainty in general, while within uncertainty, random sampling error dominates measurement error.

As promised, we now revisit the question of whether to incorporate an estimate of correlation among the various produce categories in PCB concentration and in consumption rate. Because two inputs, C_v and M_v, are substantially more important than any others for this model, a very high degree of correlation between the various produce categories would have to exist in order for correlation to make an important difference in the results (Smith et al., 1992). A correlation between C_v and M_v would potentially affect our results; however, these inputs are uncorrelated as the amount of produce an individual eats is not expected to be associated with the trace levels of contamination. Thus, correlation remains unaccounted for in the analysis.

The results described above provide a starting point for thinking about subsequent analyses and the value of collecting additional information. For example, resources directed toward reducing random sampling error by increasing sample sizes would alter the results more than would improving measurement technique and reducing measurement error. On the other hand, variability, the irreducible component, is the most significant contributor to overall variance in dose. For the pathways considered in this case, variability is mostly dominated by house-to-house heterogeneity in indoor air concentrations. Thus, we see that information about the indoor concentrations of PCBs in individual homes provides the best indicator of which variability percentile represents a particular individual's exposure or dose. An understanding of the source of the indoor air contamination would be most informative in ultimately reducing these exposures.

GLOSSARY

Abscissa   The horizontal axis of a graph.

Accuracy   The measure of agreement between model predictions and measured observations of a quantity.

Arithmetic mean   The sum of all the measurements in a data set divided by the number of measurements in the data set.

Assessment endpoint   The quantity required as the output of an exposure assessment. This depends on the nature of the policy or compliance problem which the assessment is intended to address. For carcinogenic assessment end points, the appropriate exposure descriptor is a long-term (usually lifetime) average exposure. For acute health effect endpoints, a peak exposure or short-term average is usually required.

Bayesian approach (or "subjectivist" approach)   In the subjectivist or Bayesian approach, the assessment of the probability of an outcome is based on a "degree of belief" that the outcome will occur, based on all of the relevant information an analyst currently has about the system. Under the Bayesian approach a "prior" distribution is constructed that expresses one's state of knowledge (or ignorance) about the parameters of a probability model. The prior distribution can be developed on purely subjective grounds in the absence of data. The prior judgment is then "updated" using additional judgments or data from other sources, yielding a "posterior" distribution which is shaped by all of the information used in its generation. See Section 2.1.2 for more discussion.

Bias   Systematic error. For example, bias (or systematic error) in a measurement technique will keep the mean of a measured quantity from converging to the "true" mean value of that quantity.

Bimodal   A distribution that contains two *modes*. A mode is a peak on *a probability density function*.

Bounding estimate   An estimate of exposure, dose, or risk that is higher than that incurred by the person in the population who has the highest exposure, dose, or risk. Bounding estimates are useful in developing statements that exposures, doses, or risks are "not greater than" the estimated value.

Censored data   Empirical measurements that are restricted in the range of values they can assume either by a feature of the method of sampling and analysis, or by post hoc excision of values outside of a specified range.

Coefficient of variation   For a distribution or data set, the coefficient of variation is the standard deviation divided by the mean. This quantity is also known as relative standard deviation.

Comparability   The ability to describe likenesses and differences in the quality and relevance of two or more data sets.

Complementary cumulative distribution function   The function which gives the probability that a random variable takes on a value greater than some specified value.

Confidence interval   A range of numbers believed to include an unknown quantity with a specified confidence. Associated with the interval is a measure of the confidence we have that the interval does indeed contain the quantity.

Continuous random variable   A random variable that may take on any real value in an interval of numbers (its possible values are infinite).

Correlation coefficients   For a probabilistic analysis, correlation coefficients provide an estimate of the magnitude of linear influence (importance) of uncertainty in a model input to uncertainty in a model output.

Cumulative distribution function (CDF)   Obtained by integrating the Probability Density Function (PDF). Provides a quantitative relationship between the value of a quantity and the cumulative probability (percentile) of that quantity.

Data quality objectives (DQO)   Qualitative and quantitative statements of the overall level of uncertainty that a decision-maker is willing to accept in results or decisions derived from environmental data. DQOs provide the statistical framework for planning and managing environmental data operations consistent with the data user's needs.

Decision variables   These are quantities over which a decision-maker exercises control, such as the maximum acceptable emission rate for a given source. A decision-maker selects this value. Thus it is not appropriate to treat it probabilistically (see Section 2.2.3).

Defined constants   Quantities whose values are defined by convention and are not uncertain, e.g., π (pi). See Section 2.2.3.

Deterministic exposure analysis   An exposure analysis in which no information on model or input uncertainty is accounted for quantitatively. The purpose of a deterministic analysis is to provide decision-makers with a point-estimate that can be used in comparison with other analyses of exposure.

Distribution See *Probability distribution*

Dose-response relationship   The resulting biological responses in an organ or organism expressed as a function of a series of different doses.

Dose The amount of a substance available for interaction with metabolic processes or biologically significant receptors after crossing the outer boundary of an organism. The potential dose is the amount ingested, inhaled, or applied to the skin. The applied dose is the amount of a substance presented to an absorption barrier and available for absorption (although not necessarily having yet crossed the outer boundary of the organism). The absorbed dose is the amount crossing a specific absorption barrier (e.g., the exchange boundaries of skin , lung, and digestive tract) through uptake processes. Internal dose is a more general term denoting the amount absorbed without respect to specific absorption barriers or exchange boundaries. The amount of the chemical available for interaction by any particular organ or cell is termed the delivered dose for that organ or cell.

Empirical Measurable, at least in principle, e.g., pollutant concentration.

Expert judgment Judgment by an expert. Judgment involves a reasoned formation of opinions. An expert is someone with special knowledge or experience in a particular problem domain. Expert judgment is documented and can be explained to satisfy outside scrutiny.

Exposure assessment The determination or estimation (qualitative or quantitative) of the magnitude, frequency, duration, and route of exposure.

Exposure pathway The physical course a chemical or pollutant takes from the source to the organism exposed.

Exposure scenario A set of facts, assumptions, and inferences about how exposure takes place that aids the exposure assessor in evaluating, estimating, or quantifying exposures.

Exposure Contact of a chemical, physical, or biological agent with the outer boundary of an organism. Exposure is quantified as the concentration of the agent in the medium in contact integrated over the time duration of that contact.

Fractile The value of a random variable X for a specified percentile of its distribution.

Goodness-of-fit test An approach to evaluating the potential inadequacies of a model with respect to its fitness to represent a data set.

Hazard A possible harm or adverse outcome.

Hazard assessment (or hazard identification) The identification and exploration of a hazard.

Independent A property of two events such that the likelihood of one occurring is not related in any way to whether or not the other occurs, or a property of two inputs such that the value assumed by one is not related by the value assumed by the other.

Influential input An input whose range of uncertainty contributes significantly to the range of uncertainty in a model output. In a probabilistic analysis, influential inputs may be identified as those inputs having the greatest correlation with model outputs.

Inputs   Quantities which are factors, variables or parameters in a model. See Chap. 1 for more discussion.

Kurtosis   A measure of the shape of a distribution based upon the fourth moment of the distribution. Kurtosis is an indication of the flatness or peakedness of a distribution.

Latin hypercube sampling (LHS)   A numerical technique for generating sample values from specified probability distributions of random variables. In this method the sample space of an input is divided into equiprobable intervals. Samples are drawn once from each of these intervals, without replacement, to ensure representation of a range of possible values, making this a more efficient sampling technique for most purposes than Monte Carlo simulation. See Section 7.5.2.

Mathematical model   A mathematical description of a process or system.

Maximum entropy inference (MEI)   A distribution development technique involving subjective probability assessment (see Section 6.2). The maximum entropy technique produces distributions that are very broad since no mathematical possibility is ignored unless it is precluded by the information available, and with minimum bias.

Mean   The probability-weighted average of a set of values or a probability distribution. Also referred to as the "expected value" or the first moment of a probability distribution. It is a mathematical operator with important properties with respect to the addition of two or more distributions or the multiplication of a distribution by a constant. See Sections 5.1.1.1 and 7.2.

Median value   The value in a set of measurements or probability density function such that half the measured values are greater and half are less. For even numbers of data points the average of the two points lying in the central position is the median.

Mode   The value or values in a data set (or probability density function) that occur most frequently. Distributions with more than one mode are referred to as multimodal. See Section 5.1.1.1.

Model domain parameters   These are parameters that are associated with a model but not directly with the phenomenon that the model represents, e.g., spatial or temporal grid size in a numerical model. See Section 2.2.3.

Model validation   A process of hypothesis testing structured to determine whether a model can be rejected as false. Model validation is the comparison of model results to observations from the system which is being modeled. See Section 3.5.

Model verification   A process of ensuring that the mathematical structure of a model, its computer implementation, and its input assumptions are as intended. See Section 3.4.

Model   A set of constraints restricting the possible joint values of several random variables. A hypothesis or system of belief about how a system works or responds to changes in its inputs. The purpose of a model is to represent as accurately as *necessary* a system of interest. See Chapter 3.

Moments   A probability distribution may be characterized by moments of various orders. The first and second moments are the mean and variance, respectively. The third moment is used to calculate skewness. The fourth moment is used to calculate kurtosis. See Section 5.1.

Monte Carlo simulation or technique   A numerical random sampling method for generating a representative distribution of values for each of the inputs in a generic exposure (or dose) equation to derive an output distribution of exposures (or doses) in the population.

Outputs   Quantities which result from solution of an exposure assessment model.

Overconfidence   A tendency to underestimate the level of uncertainty associated with judgments or measurement.

Parameter   A quantity used to calibrate or specify a model, such as parameters of a probability model, e.g., mean and variance of a normal distribution, or slope and intercept of a least squares univariate regression model. Parameter values are usually selected by fitting the model to a calibration data set.

Parametric distribution   A probability model that is specified by parameters. For example, the normal distribution is a parametric distribution. The parameters of the normal distribution are the mean and standard deviation.

Partial correlation   The relationship between a model input and output, isolated from the impact of other inputs. See Section 8.3.2.

Pathway   The physical course or route that a pollutant takes between a source and a receptor.

Percentile   The value associated with the probability that a random variable will assume values less than or equal to a given fractile.

Precision   A measure of the reproducibility of a measured value under a given set of conditions.

Probabilistic analysis   Analysis in which frequency (or probability) distributions are assigned to represent variability (or uncertainty) in quantities. The form of the output of probabilistic analysis is likewise a distribution.

Probability density function (PDF)   Graphical or tabular representation of the relative likelihoods with which an unknown or variable quantity may obtain various values. The sum (or integral) of all likelihoods must equal one for discrete (continuous) random variables. See Section 4.3.

Probability distribution   The mathematical description of the function relating probabilities with specified intervals of values, for a random variable.

Probability models   An analytical model for a probability distribution, which is typically expressed as a probability density function.

Probability paper   Graph paper with the *x*-axis scaled in units of standard deviation or with other transformations. On this paper the cumulative distribution function for a normally distributed set of data will appear as a straight line.

Probability samples   Samples selected from a population such that each sample has a known probability of being selected.

Random numbers   A set of numbers for which no pattern or model explains the sequence.

Random samples   Samples selected from a population such that each sample has an equal probability of being selected. See also *Sample*.

Random variable   An uncertain quantity whose value depends on chance. The likelihood of selecting any particular outcome is specified by a probability model.

Range   The difference between the largest and smallest values in a measurement data set.

Rank correlation   A measure of the linear dependence of the ranks of values generated to represent a particular input and the model output which employs values of the input in its estimation. See Section 8.3.2.

Receptor   A human (or other) individual for whom exposure, i.e., contact under a defined set of conditions to a contaminant of concern, is assessed.

Relative Standard Deviation   The standard deviation of a random variable divided by its mean. Also called the *Coefficient of Variation*. See Section 8.2.2.

Representativeness   The property of a sample, or samples, that they are characteristic of the whole medium, exposure, or dose which they represent, and thus appropriate to serve as the basis for making inferences.

Risk characterization   The description of the nature and often the magnitude of human or nonhuman risk, including attendant uncertainty.

Risk   The probability of deleterious health or environmental effects.

Robust   A property of a decision, result or estimate such that it holds across a range of scenarios and conditions. For example, if the results from competing models lead to similar decisions, then the decision may be called robust in the face of alternative theories.

Sample   A single observation or value of a random variable. Also referred to as a sample point. See also *Random Sample*.

Sample space   Set of all possible outcomes of a function.

Scenario   See *Exposure Scenario*.

Screening analysis   An approach which uses health-conservative default assumptions in the representation of variability/uncertainty in parameters and modeling. This approach is useful for identifying low risk situations that do not require further attention.

Sensitivity analysis   The assessment of the impact of changes in input values or assumptions on generated output values. See Section 3.1.

Simulation   A series of calculations that attempt to predict a value or outcome. Simulation is useful for exploring outcomes that are not observable under the conditions of interest, or outcomes whose occurrence is potentially dangerous. See Section 7.5.

Skewness   A measure of the shape of a probability distribution. A symmetric distribution has zero skewness. A distribution with a long tail to the left (toward large negative values) is negatively skewed. A distribution with a long tail to the right (toward large positive values) is positively skewed. See Section 5.1.3.

Statistical significance   The satisfaction of a statistical criterion dictating the allowable probability of wrongly rejecting a true hypothesis. The inference that an observed result is statistically significant is typically based on a test to reject one hypothesis and accept another.

Stochastic process   A random process, a process not explainable by mechanistic theory. Sometimes processes appear to be stochastic only because of the lack of an available theory to explain them.

Stratification   Division (of a quantity) into intervals, categories, or strata.

Subjective probability distribution   A probability distribution that represents an individual's (or group's) belief about the range and likelihood of values for a quantity, based on that person's experience and knowledge about the quantity or similar quantities.

Surrogate data   Substitute data or measurements on one quantity used to estimate analogous or corresponding values of another quantity.

Tolerance interval   An interval within which a statement of confidence is given that at least a prespecified proportion of values from a sampled distribution is located. Tolerance intervals are used when the parameters of a distribution are not known precisely, but instead are estimated based upon representative data sets with small sample sizes. Thus, tolerance intervals account for the sampling error associated with making estimates of probability ranges for random quantities.

Truncated data   A set of data in which values outside of a specified range do not appear (i.e., are not measured or not reported).

Uncertainty analysis   Method of quantifying and qualifying variability/uncertainty in model outputs using probability distributions. Using any of a variety of techniques (see Chap. 7), simultaneous variabilities and uncertainties in any number of model inputs may be propagated through a model to determine their combined effect on model outputs. The results of an uncertainty analysis include the range and relative likelihood of values for the outputs. Uncertainty analysis sometimes also refers to sensitivity analysis performed to gain insights into key sources of uncertainty. See Chap. 8.

Uncertainty   Lack of knowledge about the "true" value of a quantity, lack of knowledge about which of several alternative model representations best describes a biological/chemical/physical/other mechanism of interest, or lack of knowledge about which of several alternative probability density functions should represent a quantity of interest. Uncertainty is a property of the analyst.

Value parameters   Value parameters represent the preferences or value judgments of a decision maker. Examples include the discount rate and parameters of utility functions used in decision analysis.

Variability   Heterogeneity of values over time, space, or different members of a population. Variability is a property of nature.

Variable   Adjective: a quantity that is subject to variability. Noun: a quantity that may take on different values. Note: to avoid causing confusion we avoid the use of the term variable preferring "exposure model input," except in discussion of random variables in general.

Worst case   A semiquantitative term referring to the maximum possible exposure, dose, or risk, that can conceivably occur, whether or not this exposure, dose, or risk actually occurs or is observed in a specific population. Historically this term has been loosely defined in an ad hoc way in the literature, so assessors are cautioned to look for contextual definitions when encountering this term. It should refer to a hypothetical situation in which everything that can plausibly happen to maximize exposure, dose, or risk does in fact happen. This worst case may occur (or even be observed) in a given population, but since it is usually a very unlikely set of circumstances, in most cases, a worst-case estimate will be somewhat higher than occurs in a specific population. As in other fields, the worst-case scenario is a useful device when low probability events may result in a catastrophe that must be avoided even at great cost, but in most health risk assessments, a worst-case scenario is essentially a type of bounding estimate.

REFERENCES

Abramowitz, M. and Stegun, I.A., 1965, *Handbook of Mathematical Functions,* Dover Publications, Inc., New York.

AIHC (American Industrial Health Council), 1994, Exposure Factors Sourcebook, Suite 760, 2001 Pennsylvania Avenue, NW, Washington, DC, 20006–1807.

Aitcheson, J. and Brown, J.A.C., 1957, *The Lognormal Distribution,* Cambridge University Press, Cambridge, UK.

American Statistical Association (ASA), 1992, Contemporary Statistics, Volume 1: *Combining Information, Statistical Issues and Opportunities for Research,* National Academy Press, Washington, DC.

Ang, and Tang, W.H., 1975, *Probability Concepts in Engineering Planning and Design, Volume 1.* John Wiley and Sons, New York.

Ang, and Tang, W.H., 1984, *Probability Concepts in Engineering Planning and Design, Volume 2: Decision, Risk, and Reliability.* John Wiley and Sons, New York.

Barry, T.M., 1996, Recommendations on the testing and use of pseudo-random number generators used in Monte Carlo analysis for risk assessment, *Risk Assessment,* 16(1):93–105.

Bartell, S.M., and Wittrup, M.B., 1996, The McArthur River ecological risk assessment: A case study. In: *Risk Assessment and Management Handbook; For Environmental, Health and Safety Professionals,* (R. Kolluru, S. Bartell, R. Pitblado, and S. Stricoff, eds.), McGraw Hill, Inc., New York.

Belcher, G.D. and Travis, C.C.,1991, An uncertainty analysis of food chain exposure to pollutant emitted from municipal waste combustors. In: *Health Effects of Municipal Waste Incineration,* (H. Hattemer-Frey, and C. Travis, eds.), CRC Press, Boca Raton.

Benjamin, J.R. and Cornell, C.A., 1970, *Probability and Statistics and Decisions for Civil Engineers,* McGraw-Hill, New York, NY.

Bevington, P.R., and Robinson, D.K., 1992, *Data Reduction and Error Analysis for the Physical Sciences,* McGraw-Hill, New York.

Blom, G., 1958, *Statistical Estimates and Transformed Beta-Variables,* John Wiley and Sons, New York.

Bobee, B.B. and Robitaille, R., 1975, Correction of bias in estimation of the coefficient of skewness, *Water Resources Research,* 11(6):851–854.

Bogen, K.T., 1992, RiskQ: An Interactive Approach to Probability, Uncertainty, and Statistics for Use with Mathematica (reference manual), Lawrence Livermore National Laboratory, Livermore, CA.

Bogen, K.T., 1990, *Uncertainty in Environmental Health Risk Assessment,* Garland Publishing Inc., New York.

Bogen, K.T. and Spear, R.C., 1987, Integrating uncertainty and variability in environmental risk assessment, *Risk Analysis*, 7:427–436.

Bowman, K.O. and Shenton, L. R., 1986, Moment techniques. In: *Goodness-of-Fit Techniques* (R.B. D'Agostino and M.A. Stephens, eds.), Marcel Dekker, New York.

Box, G.E.P. and Tiao, G.C., 1973, *Bayesian Inference in Statistical Analysis*, Wiley-Interscience, New York.

Brainard, J. and Burmaster, D.E., 1992, Bivariate distributions for height and weight of men and women in the United States, *Risk Analysis*, 12: 267–275.

Brattin, W.J., Barry, T.M., and Chiu, N., 1996, Monte Carlo modeling with uncertain probability density functions, *Human and Ecological Risk Assessment*, 2(4):820–840.

Broadbent, S.R., 1956, Lognormal approximation to products and quotients, *Biometrica*, 43:404–417.

Buckley, J.J., 1985, Entropy principles in decision making under risk, *Risk Analysis*, 5:303–313.

Bukowski, J.L., Korn, L., and Wartenberg, D., 1995, Correlated inputs in quantitative risk analysis: The effect of distributional shape, *Risk Analysis*, 15:215–219.

Burmaster, D.E. and Harris, R.H., 1993, The magnitude of compounding conservatisms in Superfund risk assessments, *Risk Analysis*, 13(2):131–143.

Burmaster, D.E. and Thompson, K. M., 1995, BackCalculating cleanup targets in probabilistic risk assessments when the acceptability of cancer risk is defined under different risk management policies, *Human and Ecological Risk Assessment*, 1(1):101–120.

Burmaster, D.E. and vonStackelberg, K., 1991, Using Monte Carlo simulations in public health risk assessments: Estimating and presenting full distributions of risk, *JEAEE*, 1:491–512.

Burmaster, D.E. and Wilson, A.M., 1996, An introduction to second-order random variables in human health risk assessments, *Human and Ecological Risk Assessment*, 2(4):892–919.

Carrington, C.D., 1993, How can you reduce uncertainty if you can't measure it?, Presented at 1993 Symposium on Biological Mechanisms and Quantitative Risk Assessment, U.S. Environmental Protection Agency, Research Triangle Park, NC, November 1–4.

Chatterjee, S. and Price, B., 1991, *Regression Analysis by Example*, 2nd ed., John Wiley and Sons, New York.

Cmelik, R.F. and Gehani, N.H., 1988, Dimensional analysis with C++, *IEEE Software*, 5(3):21–27.

Cohen, J.T., Lampson, M.A., and Bowers T.S., 1996, The use of two-stage Monte Carlo simulation techniques to characterize variability and uncertainty in risk analysis, *Human and Ecological Risk Assessment*, 2(4):939–971.

Constantinou, E., Seigneur, C., and Permutt T., 1992, Uncertainty analysis of health risk estimates: Application to power plant emissions, Paper 92-95.01, *Proceedings of the 85th Annual Meeting*, Air and Waste Management Association, Pittsburgh, PA, June, 1992.

Cooke, R.M., 1991, *Experts in Uncertainty: Opinion and Subjective Probability in Science*, Oxford Press, New York.

Cox, D.R., 1962, *Renewal Theory*, Methuen and Company, London, UK.

Cox, D.R. and Oakes, D., 1984, *Analysis of Survival Data*, Chapman and Hall, London, UK.

Crouch, E.A.C. and Wilson, R., 1981, Regulation of carcinogens, *Risk Analysis*, 1:47–57.

Crowder, M.J., Kimber, A.C., Smith, R.L., and Sweeting, T.J., 1991, *Statistical Analysis of Reliability Data*, Chapman and Hall, London, UK.

Cullen, A.C., 1994, Measures of conservatism in probabilistic risk assessment, *Risk Analysis*, 14: 389–393.

Cullen, A.C., 1995, The sensitivity of Monte Carlo simulation results to model assumptions: The case of municipal solid waste combustor risk assessment, *J Air Waste Mgmt Assoc*, 45:538–546.

Cullen, A.C., Vorhees, D.V., and Altshul, L.M., 1996, The influence of harbor contamination on the level and composition of polychlorinated biphenyls (PCBs) in produce in greater New Bedford, MA, *ES&T*, 30:1581–1588.

D'Agostino, R.B., 1986, Graphical analysis. In: *Goodness-of-Fit Techniques* (R.B. D'Agostino and M.A. Stephens, eds.), Marcel Dekker, New York, pp. 7–62.

D'Agostino, R.B. and Stephens, M.A. (eds.), 1986, *Goodness-of-Fit Techniques*, Marcel Dekker, New York.

Decisioneering, 1996, *Crystal Ball Version 4.0 User Manual*, Decisioneering, Inc., Denver, CO.

DeGroot, M.H., 1969, *Optimal Statistical Decisions*, McGraw–Hill Book Company, New York.

DeGroot, M.H., 1986, *Probability and Statistics*, 2nd ed., Addison–Wesley Pub. Inc., Reading, MA.

Dillon, W. and Goldstein, M., 1984, *Multivariate Analysis: Methods and Applications*, John Wiley and Sons, New York.

Di Toro, D.M. and Small, M.J., 1979, Stormwater interception and storage, *J Environ Eng Div ASCE*, 105(1):43–54.

Draper, N.R. and Smith, H., 1981, *Applied Regression Analysis*, 2nd ed., John Wiley and Sons, New York.

Eagleson, P.S., 1978, Climate, soil and vegetation, *Water Resources Research*, 14:705–776.

Edwards, A., 1984, *An Introduction to Linear Regression and Correlation*, 2nd ed., W.H. Freeman, New York.

Efron, B. and Tibshirani, R.J., 1993, *An Introduction to the Bootstrap*, Chapman and Hall, New York.

Enfield, D.B. and Cid S., Luis, 1991, Low-frequency changes in El Nino-southern oscillation, *Journal of Climate*, 4:1137–1146.

Evans, J., Hawkins, N., and Graham, J., 1988, The value of monitoring for radon in the home: A decision analysis, *J Air Pollut Control Assoc*, 38(11):1380–1385.

Evans, J.S., 1985, The value of improved exposure estimates: A decision analytic approach, *Proc. of the 78th Ann. Mtg. of the Air Pollution Control Association*, (June 16–21, 1985), Detroit, Michigan.

Evans, J.S., Graham, J.D., Gray, G.M., and Sielken, R.L., 1994a, A distributional approach to characterizing low-dose cancer risk, *Risk Analysis*, 14(1):25–34.

Evans, J.S., Gray, G.M., Sielken, R.L., Smith, A.E., Valdez-Flores, C., and Graham, J.D., 1994b, Use of probabilistic expert judgment in distributional analysis of carcinogenic potency, *Reg Tox and Pharm*, 20:15–36.

Ferson, S. and Long, T.F., 1995, Conservative uncertainty propagation in environmental risk assessments. In: *Environmental Toxicology and Risk Assessment*, third volume, STRM STP 1218, (J.S. Hughes, G.R. Biddinger, and E. Mones, eds.), American Society for Testing and Materials, Philadelphia, PA.

Finkel A.M. and Evans, J.S., 1987, Evaluating the benefits of uncertainty reduction in environmental health risk management, *J Air Pollut Control Assoc*, 37:1164–1171.

Finkel, A.M., 1989, Is risk assessment really too conservative? Revising the revisionists, *Columbia Journal of Environmental Law*, 14:427–467.

Finkel, A.M., 1990, *Confronting Uncertainty in Risk Management: A Guide for Decision Makers*, Center for Risk Management, Resources for the Future, Washington DC.

Finley, B.L. and Paustenbach, D.J., 1994, The benefits of probabilistic exposure assessment: Three case studies involving contaminated air, water, and soil, *Risk Anal* 14(1):53–73.

Frey, H.C., 1992, *Quantitative Analysis of Uncertainty and Variability in Environmental Policy Making*, American Association for the Advancement of Science, Washington, DC, (September 1992).

Frey, H.C., 1993, Performance model of the fluidized bed copper oxide process for SO_2/NO_x control, paper No. 93-WA-79.01, *Proceedings of the 86th Annual Meeting*, Air and Waste Management Association, Pittsburgh, PA (June, 1992).

Frey, H.C., 1997, Bootstrap methods for quantitative analysis of variability and uncertainty in exposure and risk assessment, paper No. 97-100B.03, *Proceedings, Annual Meeting of the Air & Waste Management Association*, Pittsburgh, PA (June, 1992).

Frey, H.C., 1998, Quantitative analysis of variability and uncertainty in energy and environmental systems. In: *Model and Analysis of Uncertainty in Civil Engineering* (B. Ayyub, ed.), CRC Press, Boca Raton, FL, pp. 381–423.

Frey, H.C., 1998b, Methods for the quantitative analysis of variability and uncertainty in hazardous air pollutant emissions, paper no. 98-105 B.01, *Proceedings 91st Annual Meeting of the Air and Waste Management Association*, Pittsburgh, PA (June 14-18).

Frey, H.C. and Agarwal, P., 1996, Probabilistic analysis and optimization of clean coal technologies, paper no. 96-119.02, *Proceedings 89th Annual Meeting of the Air and Waste Management Association*, Pittsburgh, PA (June 1996).

Frey, H.C. and Bharvirkar, R., 1998, Desktop modeling of the performance, emissions, and cost of gasification systems, *Proceedings, 91st Annual Meeting of the Air and Waste Management Association*, Pittsburgh, PA (June 14–18).

Frey, H.C. and Burmaster, D.E., 1998, Methods for characterizing variability and uncertainty: Comparison of bootstrap simulation and likelihood-based approaches, *Risk Analysis* (accepted January 1998 for publication).

Frey, H.C. and Rhodes, D. S., 1996, Characterizing, simulating, and analyzing variability and uncertainty: An illustration of methods using an air toxics emissions example, *Human and Ecological Risk Assessment*, 2(4):762–797.

Frey, H.C. and Rhodes, D. S., 1998, Characterization and simulation of uncertain frequency distributions: Effects of distribution choice, variability, uncertainty, and parameter dependence, *Human and Ecological Risk Assessment* 4(2):423–468.

Frey, H.C. and Rubin, E.S., 1991, Probabilistic evaluation of advanced SO_2/NO_x control technology, *Journal of the Air and Waste Management Association*, 41(12):1585–1593.

Frey, H.C. and Rubin, E.S., 1992a, Evaluation of advanced coal gasification combined-cycle systems under uncertainty, *Industrial and Engineering Chemistry Research*, 31(5):1299–1307.

Frey, H.C. and Rubin, E.S., 1992b, An evaluation method for advanced acid rain compliance technology, *Journal of Energy Engineering*, 118(1):38–55.

Frey, H.C. and Rubin, E.S., 1992c, Integration of coal utilization and environmental control in integrated gasification combined cycle systems, *Env Sci and Tech*, 26(10):1982–1990.

Gan, F.F., and Koehler, K.J., 1990, Goodness-of-fit tests based on P–P probability plots, *Technometrics*, 32(3):289–303.

Gan, F.F. Koehler, K.J., and Thompson, J.C., 1991, Probability plots and distribution curves for assessing the fit of probability models, *The American Statistician*, 45(1):14–21.

Gardner, R.H., O'Neill, R.V., Mankin, J.B., and Kumar, D., 1980, Comparative error analysis of six predator-prey models, *Ecology*, 61(2):323–332.

Glaeser, H., Hofer, E., Kloos, M., and Skorek, T., 1994, Uncertainty and sensitivity analysis of a post-experiment calculation in thermal hydraulics, *Reliability Engineering and System Safety*, 45:19–33.

Gumbel, E.J., 1958, *Statistics of Extremes*, Columbia University Press, New York.

Hahn, G.J. and Meeker, W.Q., 1991, *Statistical Intervals: A Guide for Practitioners*, John Wiley and Sons, New York.

Hahn, G.J., and Shapiro, S.S., 1967, *Statistical Models in Engineering*, Wiley Classics Library, John Wiley and Sons, New York.

Hammersley, J.M. and Handscomb, D.C, 1964, *Monte Carlo Methods*, London: Methuen and Co., Ltd., New York: John Wiley and Sons, Inc.

Hammonds, J.S., Hoffman, F.O., and Bartell, S.M., 1994, An Introductory Guide to Uncertainty Analysis in Environmental Health and Risk Assessment, SENES Oak Ridge, Inc., Oak Ridge, TN, April.

Hanna, S.R., J.C. Chang, and M.E. Fernau, 1998, Monte Carlo estimates of uncertainties in predictions by a photochemical grid model (UAM-IV) due to uncertainties in input variables, *Atmospheric Environment* (in press).

Harr, M.E., 1987, *Reliability-Based Design in Civil Engineering*, McGraw–Hill, Inc., New York.

Harter, L.H., 1984, Another look at plotting positions, *Commun Statist-Theor Meth*, 13(13):1613–1633.

Hastings, N.A.J. and Peacock, J.B., 1974, *Statistical Distributions: A Handbook for Students and Practitioners*, Butterworths, London, UK.

Hattis, D. and Burmaster, D., 1994, Assessment of variability and uncertainty distributions for practical risk assessments, *Risk Analysis*, 14(5):713–730.

Hawkins, N.C., 1991, Conservatism in maximally exposed individual (MEI) predictive exposure assessments: A first-cut analysis, *Reg Tox and Pharm*, 14:107–117.

Hazen, A., 1914, Storage to be provided in impounding reservoirs for municipal water supply, *Transactions of the American Society of Civil Engineers*, 77:1539–1640.

HCRA (Harvard Center for Risk Analysis), 1994, *A Historical Perspective on Risk Assessment in the Federal Government*, Harvard School of Public Health, Boston, MA.

Helton, J. et al., 1996, Computational implementation of a systems prioritization methodology for the waste isolation pilot plant: A preliminary example, Sandia Report, SAND94-3069, UC-721, Prepared by Sandia National Laboratories, Albuquerque, NM, 87185, for the US DOE under contract DE-AC04-94AL85000.

Helton, J., 1993, Uncertainty and sensitivity analysis techniques for use in performance assessment for radioactive waste disposal, *Reliability Engineering and System Safety*, 42:327–367.

Helton, J., 1996, Probability, conditional probability and complementary cumulative distribution functions in performance assessment for radioactive waste disposal, contractor report SAND95-2571, UC-721, Prepared by Sandia National Laboratories, Albuquerque, NM, 87185, for the US DOE under contract DE-AC04-94AL85000.

Henrion, M., 1982, The value of knowing how little you know: The advantages of probabilistic treatment in policy analysis, Ph.D. dissertation, Carnegie-Mellon University.

Henrion, M., and Fischhoff, B., 1986, Assessing uncertainty in physical constants, *Am J Phys*, 54:791–797.

Henrion, M., Marnicio, R.J., and Bloyd, C.N., 1995, The Tracking and Analysis Framework: A Tool for the Integrated Assessment of Air Pollution Controls, Lumina Decision Systems, Los Altos, CA.

Hilfinger, P.N., 1988, An Ada package for dimensional analysis, *ACM Transactions on Programming Language and Systems*, 10(2):189–203.

Hodges, J.S., 1987, Uncertainty, policy analysis, and statistics, *Stat Science*, 2:259–291.

Hoffman, F.O. and Hammonds, J.S., 1994, Propagation of uncertainty in risk assessments: The need to distinguish between uncertainty due to lack of knowledge and uncertainty due to variability, *Risk Analysis*, 14:707–712.

Hoffman, F.O. and Miller, C.W., 1983, Uncertainties in environmental radiological assessment models and their implications, *Proceedings of the Nineteenth Annual Meeting of the National Council on Radiation Protection and Measurements*, National Academy of Sciences, Washington, DC, April 6–7.

Hoffman, F.O. and Thiessen, K.M., 1996, The use of Chernobyl data to test predictions for interindividual variability of ^{137}Cs concentrations in humans, *Reliability Engineering and System Safety*, 54(2–3):197–202.

Holland, D.M., and Fitz-Simons, T., 1982, Fitting statistical distributions to air quality data by the maximum likelihood method, *Atmospheric Environment*, 16(5):1071–1076.

Hora S.C. and Iman, R.L., 1989, Expert opinion in risk analysis: The NUREG-1150 methodology, *Nuclear Science and Engineering*, 102:323–331.

Hornbeck, R.W., (1975), *Numerical Methods*. Prentice Hall/Quantam: Edgewood Cliffs, NJ.

IAEA (International Atomic Energy Agency), 1989, Evaluating the reliability of predictions made using environmental transfer models, *Safety Series*, No. 100, Vienna, Austria.

ICRP (International Commission on Radiological Protection), 1975, Report of the Task Group on Reference Man, International Commission on Radiological Protection Number 23, Pergamon Press, Oxford, UK.

Iman, R. L. and Conover, W. J., 1982, A distribution-free approach to inducing rank correlation among input variables, *Communications in Statistics*, B11(3):311–334.

Iman, R.L. and Helton, J.C., 1988, An investigation of uncertainty and sensitivity analysis techniques for computer models, *Risk Analysis*, 8(1):71–90.

Iman, R.L. and Hora, S.C., 1989, Bayesian methods for modeling recovery times with an application to the loss of off-site power at nuclear power plant, *Risk Analysis*, 9(1):25–36.

Iman, R.L. and Hora, S.C., 1990, A robust measure of uncertainty importance for use in fault tree system analysis, *Risk Analysis*, 10(3):401–406.

Iman, R.L. and Shortencarier, M.J., 1984, A Fortran 77 program and user's guide for the generation of Latin hypercube and random samples for use with computer models, SAND83-2365, Sandia National Laboratory, Albuquerque, NM, (January, 1984).

Iman, R.L., Shortencarier, M.J., and Johnson, J.D., 1985, A Fortran 77 program and user's guide for the calculation of partial correlation and standardized regression coefficients, SAND85-0044, Sandia National Laboratory, Albuquerque, NM, (June 1985).

Ishigami, T., Cazzoli, E., Khatig-Rahbar, M., and Unwin, S.D., 1989, Techniques to quantify the sensitivity of deterministic model uncertainties, *Nuclear Science and Engineering*, 101:371–383.

Israeli, M. and Nelson, C.B., 1992, Distributions and expected time of residence for U.S. households, *Risk Analysis*, 12:65–72.

Jablon, S., 1993, Neutrons in Hiroshima and uncertainties in cancer risk estimates, *Radiation Research*, 133:130–131.

Jaynes, E.T., 1957, Information theory and statistical mechanics, *Phys Rev*, 106:620–630.

Johnson, N.L., 1980, Fitting distributions to data, ASQC Chemical Division Technical Conference, Cincinnati, Ohio, October.

Johnson, N.L. and Kotz, S., 1970, *Continuous Univariate Distributions*, Vol. 2, Houghton Mifflin Company, Boston.

Johnson, T., Capel, J., McCoy, M., 1993, Estimating of ozone exposures experienced by urban Residents using a probabilistic version of NEM and 1990 population data, Prepared by International Technology Air Quality Services for U.S. Environmental Protection Agency (September).

Justus, C.G., Hargraves W.R., Yalcin, A., 1976, Nationwide assessment of potential output from wind powered generators, *J Appl Meteorology* 15:673–678.

Kahneman, D., Slovic, P., and Tversky, A. (eds.), 1982, *Judgment Under Uncertainty: Heuristics and Biases*, Cambridge University Press, New York.

Kalagnanam, J., Kandlikar, M., and Linville, C., 1998, Importance ranking of model uncertainties: A robust approach, under review at *Risk Analysis*.

Kaplan, P.G., 1991, A formalism to generate probability distributions for performance assessment modeling, *Proceedings of the Second Annual International Conference on High Level Radioactive Waste Management*, April 28–May 3, Las Vegas, NV.

Kaplan, S. and Garrick, B.J., 1981, On the quantitative definition of risk, *Risk Analysis*, 1:11–27

Kapur, J. and Kasavan, H., 1992, *Entropy Optimization Principles with Applications*, Academic Press Inc., Boston, MA.

Kini, M.D. and Frey, H.C., 1997, Probabilistic Evaluation of Mobile Source Air Pollution: Volume 1, Probabilistic Modeling of Exhaust Emissions from Light Duty Gasoline Vehicles, Prepared by North Carolina State University for Center for Transportation and the Environment, Raleigh, December 1997.

Kirchner, T.B., 1990, Establishing modeling credibility involves more than validation, *Proceedings, On the Validity of Environmental Transfer Models*, Biospheric Model Validation Study, Stockholm, Sweden, October 8–10.

Kite, G.W., 1977, *Frequency and Risk Analysis in Hydrology*, Water Resources Publications, Littleton, Colorado.

Klee, A.J., 1992, MOUSE: A Computerized Uncertainty Analysis System Operational Manual, U.S. Environmental Protection Agency, Cincinnati, OH, July.

Laha, R.G. and Rohatgi, V.K., 1979, *Probability Theory*, John Wiley and Sons, New York.

Lambert, J., Matalas, N., Ling, C., Haimes, Y., and Li, D. 1994, Selection of probability distributions in characterizing risk of extreme events, *Risk Analysis*, 14:731–742.

Lave, L.B., Ennever, F.K., Rosenkranz, H.S., and Omenn, G.S., 1988, Information value of the rodent bioassay, *Nature*, 336:631–633.

Law, A.M., and Kelton, W.D., 1991, *Simulation Modeling and Analysis*, 2nd ed., McGraw-Hill, New York.

Layton, D., 1993, Metabolically consistent breathing rates for use in dose assessments, *Health Physics*, 64:23–36.

Lee, E.T., 1992, *Statistical Methods for Survival Data Analysis*, 2nd ed., John Wiley and Sons, New York.

Lee, R.C. and Wright, W.E., 1994, Development of human exposure-factor distributions using maximum-entropy inference, *J of Exp Anal and Env Epi*, 4(3):329–341.

Lee, R.C., Fricke, J.R., Wright, W.E., and Haerer, W., 1995, Development of a probabilistic blood lead prediction model, *Environmental Geochemistry and Health*, 17: 169–181.

Lettenmaier, D.P. and Burges, S.J., 1982, Gumbel's extreme value I distribution: A new look, *J Hydraulics Div, ASCE*, 108(4): 502–514.

Levine, R.D. and Tribus, M., 1978, *The Maximum Entropy Formalism*, MIT Press, Cambridge, MA.

Lumina, 1996, *Analytica User Guide*, Lumina Decision Systems, Los Altos, CA.

MacIntosh, D.L., Suter II, G.W., and Hoffman, F.O., 1994, Uses of probabilistic exposure models in ecological risk assessments of contaminated sites, *Risk Analysis*, 14(4):405–419.

Mandel, J., 1969, *The Statistical Analysis of Experimental Data*, John Wiley and Sons, New York.

March, J.G. and Simon, H.A., 1958, *Organizations*, John Wiley and Sons, New York.

McKay, M.D., Conover, W.J., and Beckman, R.J., 1979, A comparison of three methods for selecting values of input variables in the analysis of output from a computer code, *Technometrics*, 21(2):239–245.

McKone, T.E., 1996, The reliability of a three-compartment fugacity model for estimating the long-term inventory and flux of contaminants in soils, *Reliability Engineering and Systems Safety*, 54:165–181.

McKone, T.E. and Ryan, P.B., 1989, Human exposures to chemicals through food chains: An uncertainty analysis, *ES&T*, 23(9):1154–1163.

McLachlan, G.J., and Basford, K.E., 1988, *Mixture Models: Inference and Applications to Clustering*, Marcel Dekker, New York.

Merkhofer, M.W., 1987, Quantifying judgmental uncertainty: Methodology, experiences, and insights, *IEEE Transactions on Systems, Man, and Cybernetics*. 17(5):741–752.

Morgan, M.G. and Henrion, M., 1990, *Uncertainty: A Guide to Dealing with Uncertainty in Quantitative Risk and Policy Analysis*, Cambridge University Press, New York.

Morgan, M.G., Henrion, M., and Morris, S.C., 1980, Expert Judgment for Policy Analysis, Brookhaven National Laboratory, BNL 51358 Brookhaven, NY.

Murray, D.M. and Burmaster, D.E., 1993, Review of RiskQ: An interactive approach to probability, uncertainty, and statistics for use with Mathematica™, *Risk Analysis*, 13(4):479–482.

NCRP (National Council on Radiation Protection and Measurements), 1996, A Guide for Uncertainty Analysis in Dose and Risk Assessments Related to Environmental Contamination, NCRP Commentary No. 14, Bethesda, MD.

Nelson, R.B., 1986, Properties of a one-parameter family of bivariate distributions with specified marginals, *Communications in Statistics—Theory and Method*, A15:3277–3285.

NRC, 1983, *Risk Assessment in the Federal Government: Managing the Process*, Committee on the Institutional Means for Assessment of Risks to Public Health, National Academy Press, Washington DC.

NRC, 1994, *Science and Judgment in Risk Assessment*, National Research Council, National Academy Press, Washington DC.

O'Neill, R.V., Gardner, R.H., and Mankin, J.B., 1980, Analysis of parameter error in a nonlinear model, *Ecological Modelling*, 8:297–311.

Ott, W., 1990, A physical explanation of the lognormality of pollutant concentrations, *J Air Waste Manage Assoc*, 40: 1378–1383.

Ott, W., 1995, *Environmental Statistics and Data Analysis*, Lewis Publishers, Boca Raton, FL.

Palisades, 1997, web site for @Risk, www.palisade.com.

Pate-Cornell, M.E., 1996, Uncertainties in risk analysis: Six levels of treatment, *Reliability Engineering and System Safety*, 54:95–111.

Price, P., Curry, C., Goodrum, P., Gray, M., McCrodden, J., Harrington, N., Carlson-Lynch, H., and Keenan, R., 1996a, Monte Carlo modeling of time-dependent exposures using a microexposure event approach, *Risk Analysis*, 16:339–348.

Price, P., Su, S., Harrington, J., and Keenan, R., 1996b, Uncertainty and variation in indirect exposure assessment: An analysis of exposure to tetrachlorodibenzo-p-dioxin from a beef consumption pathway, *Risk Analysis*, 16:263–277.

Raiffa, H., 1968, *Decision Analysis: Introductory Lectures on Choices Under Uncertainty*, Addison-Wesley Publishing, Reading, MA.

Raiffa, H. and Schlaifer, R.O., 1961, *Applied Statistical Decision Theory*, Harvard University Press, Cambridge, MA.

Rao, S.S., 1992, *Reliability-Based Design*, McGraw-Hill, New York, NY.

NRC, 1975, *Reactor Safety Study: An Assessment of Accident Risks in the U.S. Commercial Nuclear Power Plants*, Nuclear Regulatory Commission, NUREG-75/014 (WASH-1400) Washington, DC.

Reichard, E.G. and Evans, J.S., 1989, Assessing the value of hydrogeologic information for risk based remedial action decisions, *Water Res Res*, 75:1451–1460.

Resendiz-Carrillo, D. and Lave, L.B., 1987, Optimizing spillway capacity with an estimated distribution of floods, *Water Res Res*, 23(11): 2043–2049.

Rhodes, D.S., 1997, Quantitative analysis of variability and uncertainty in environmental risk assessment, master's thesis, Department of Civil Engineering, North Carolina State University, Raleigh, NC, May.

Rhodes, D.S. and Frey, H.C., Quantification of Variability and Uncertainty in AP-42 Emission Factors Using Bootstrap Simulation, in *Proceedings, Emission Inventory: Planning for the Future, Air & Waste Management Association*, Pittsburgh, PA, October, 1997.

Rish, W.R., 1988, Approach to Uncertainty in Risk Analysis, ORNL/TM-10746, Oak Ridge National Laboratory, Oak Ridge, TN, August.

Rubin, E.S., Small, M.J., Bloyd, C.N., and Henrion, M., 1992, Integrated assessment of acid-deposition effects on lake acidification, *J Environmental Engineering, ASCE*, 118:120–134.

Rugen, P. and Callahan, B., 1996, An overview of Monte Carlo, A fifty year perspective, *Human and Ecological Risk Assessment*, 2:671–680.

Scheuer, E.M. and Stoller, D.S., 1962, On the generation of normal random vectors, *Technometrics*, 4:278–281.

Seinfeld, J.H., 1986, *Atmospheric Chemistry and Physics of Air Pollution*. Wiley-Interscience, New York.

Sexton, K. and Ryan, P.B., 1987, Assessment of human exposure to air pollution methods, measurements, and models. In: *Air Pollution, the Automobile and Public Health: Research Opportunities for Quantifying Risk*, National Academy Press, Washington, DC.

Shannon, C.E. and Weaver, W., 1949, *The Mathematical Theory of Communication*, University of Illinois Press, Urbana, IL.

Shaw, C.D. and Burmaster, D.E., 1996, Distributions of job tenure for US workers in selected industries and occupations, *Human and Ecological Risk Assessment*, 2:798–819.

Shlyakhter, A.I. and Kammen, D.M., 1992, Sea-level rise or fall? *Nature*, 253:25.

Shlyakhter, A.I., 1994, An improved framework for uncertainty analysis: accounting for unsuspected errors, *Risk Analysis*, 14: 441–447.

Simon, H.A., 1996, *The Sciences of the Artificial, Third Edition*, MIT Press, Cambridge, MA.

Singh, K., 1997. Evaluation of the effect of uncertainties in the Mobile Source Emissions Inventory on Predicted Air Quality, Master's thesis, Department of Civil Engineering, North Carolina State University, Raleigh, NC.

Singh, K., and Frey, H.C. 1998, Probabilistic Evaluation of Mobile Source Air Pollution: Volume 2, Evaluation of the Effect of Uncertainties in Mobile Source Emissions on Predicted Ambient Air Quality, North Carolina State University for Center for Transportation and the Environment, Raleigh.

Slovic, P., Fischoff, B., and Lichtenstein, S., 1979, Rating the risks, *Environment*, 21:14–20, 36–39.

Small, M.J., 1990, Probability distributions and statistical estimation, Chapter 5. In *Uncertainty: A Guide to Dealing with Uncertainty in Quantitative Risk and Policy Analysis* (M.G. Morgan, and M. Henrion, eds) Cambridge University Press, New York.

Small, M.J. and Morgan, D.J., 1986, The relationship between a continuous-time renewal model and a discrete Markov chain model of precipitation occurrence, *Water Res Res*, 22(10):1422–1430.

Small, M.J. and Chicowicz, N.L., 1986, Risk-based design of inspection programs for detecting leaky underground storage tanks, *Proceedings of the ASCE Annual Conference on Environmental Engineering*, 372–379.

Small, M.J., Cosby, B.J., Marnicio, R.J., and Henrion, M., 1995, Joint application of an empirical and mechanistic model for regional lake acidification, *Environmental Monitoring and Assessment*, 35:113–136.

Smith, A.E, Ryan, P.B., and Evans, J.S., 1992, The effect of neglecting correlations when propagating uncertainty and estimating the population distribution of risk, *Risk Analysis*, 12:467–474.

Spetzler, C.S. and von Holstein, S., 1975, Probability encoding in decision analysis, *Management Science*, 22(3):340–358.

Steel, R.G.D. and Torrie, J.H., 1980, *Principles and Procedures of Statistic: A Biometric Approach*, Second Edition, McGraw-Hill, New York.

Stephens, M.A., 1986, Tests based on EDF statistics. In *Goodness-of-Fit Techniques* (R.B. D'Agostino and M.A. Stephens, eds.), Marcel Dekker, New York, pp. 97–194.

Stinson, P.J. and Walsh, J.E., 1965, An application of the Poisson approximation to naval aviation accident data, *Bulletin of the International Statistical Institute*, 41(1):379–380.

Taylor, A.C., 1992, Addressing the uncertainty in the estimation of environmental exposure and cancer potency, ScD Thesis, School of Public Health, Harvard University.

Taylor, A.C., 1993, Using objective and subjective information to generate distributions for probabilistic exposure assessment, *Journal of Exp An and Env Epi*, 3: 285–298.

Thompson, K.M. and Graham, J.D., 1996, Going beyond the single number: using probabilistic eisk assessment to improve risk management, *Human and Ecological Risk Assessment*, 2(4):1008–1034.

Thompson, K.M., Burmaster, D.E., and Crouch, E.A.C., 1992, Monte Carlo techniques for quantitative uncertainty analysis in public health risk assessments, *Risk Analysis*, 12:53–63.

Ulam, S., 1976, *Adventures of a Mathematician*, Scribners, New York.

USDA, 1983, Food Consumption: Households in the United States: Seasons and Year 1977-1978, Agricultural Research Service.

USEPA (U.S. Environmental Protection Agency), 1989, Exposure Factors Handbook, Office of Health and Environmental Assessment, Washington, DC, 20460, EPA 600/8-89/043.

USEPA (U.S. Environmental Protection Agency), 1992a, Guidance on Risk Characterization for Risk Managers and Risk Assessors, Memorandum from F.H. Habicht, Deputy Administrator, to Assistant and Regional Administrators, February 26, 1992.

USEPA (U.S. Environmental Protection Agency), 1992b, Guidelines for exposure assessment, *Federal Register*, 57(104):22888–22938, May 29.

USEPA (U.S. Environmental Protection Agency), 1993, Chapter 2, Mobile5 Inputs, U.S. Environmental Protection Agency, Ann Arbor, MI, October.

USEPA (U.S. Environmental Protection Agency), 1995a, Policy for Risk Characterization at the US Environmental Protection Agency, Memorandum from Carol Browner to assistant and regional administrators, March 21, 1995.

USEPA (U.S. Environmental Protection Agency), 1995b, SCREEN3 Model User's Guide, EPA-454/ B-95-004, U.S. Environmental Protection Agency, Research Triangle Park, NC, September.

USEPA (U.S. Environmental Protection Agency), 1996a, Exposure Factors Handbook, Science Advisory Board Review Draft, Exposure Assessment Group, EPA/600/P-96/002B a, b, and c, Washington, DC, (August 1996).

USEPA (U.S. Environmental Protection Agency), 1996b, Summary Report for the Workshop on Monte Carlo Analysis, EPA/630/R-96/010, Risk Assessment Forum, Washington, DC.

USEPA (U.S. Environmental Protection Agency), 1996c, Requirements for Preparation, Adoption, and Submittal of Implementation Plans; Final Rule, 40 CFR Parts 51 and 52, Appendix W, Federal Register, 61(156): 41837–41894, August 12, 1996.

USEPA (U.S. Environmental Protection Agency), 1996d, Study of Hazardous Air Pollutant Emissions from Electric Utility Steam Generating Units-Interim Final Report, Volume 2, Appendices A-G, EPA-453/R-96-013b, Research Triangle Park, NC, October.

USEPA (U.S. Environmental Protection Agency), 1997a, Guiding Principles for Monte Carlo Analysis, EPA/630/R-97/001, Risk Assessment Forum. Washington, DC.

USEPA (U.S. Environmental Protection Agency), 1997b, Proposed Policy for Use of Monte Carlo Analysis in Agency Risk Assessment, Memorandum of William P. Wood, Excutive Director, Risk Assessment Forum, to Dorothy E. Patton, Executive Director of Science Policy Council (8104), January 29, 1997.

Vorhees, D.J., Cullen, A.C., and Altshul, L., 1997, Exposure to polychlorinated biphenyls in residential indoor air and outdoor air near a Superfund site, *ES&T*, 31:3612–3618.

Wallace, L., Duan, N., Zeigenfus, R., 1994, Can long-term exposure distributions be predicted from short-term measurements?, *Risk Analysis*, 14(1):75–85.

Wallace, L.A., 1995, Human exposure to environmental pollutants: a decade of experience, *Clin and Exp Allergy*, 25:4–9.

Weinberg, A.M., 1972, Science and Trans-Science, *Minerva*, 10:209–222.

Weisberg, S., 1985, *Applied Linear Regression*, John Wiley and Sons, New York.

Whitfield, R.G. and Absil, M.J.G., 1994, A Probabilistic Assessment of Health Risks Associated with Short- and Long-Term Exposure to Tropospheric Ozone, prepared by Argonne National Laboratory for U.S. Environmental Protection Agency, (February).

Whitt, W., 1976, Bivariate distributions with given marginals, *The Annals of Statistics*, 4:1280–1289.

Wilson, R. and Crouch, E.A.C., 1981, Regulation of carcinogens, *Risk Analysis*, 1(1):47–57.

Wilson, R. and Crouch, E.A.C., 1987, Risk assessment and comparisons: An introduction, *Science*, 236:267–270.

Wolfson, L.J., Kadane, J.B., and Small, M.J., 1990, Expected utility as a policy-making tool: An environmental health example, in *Bayesian Biostatistics* (D.A. Berry, and D.K. Stangl, eds.), Marcel Dekker, New York, pp. 261–277.

INDEX

Made in the USA
Lexington, KY
12 January 2010